The blood supply of the vertebral column and spinal cord in man

H. V. CROCK □ H. YOSHIZAWA

The blood supply of the vertebral column and spinal cord in man

with 120 illustrations and 44 color plates

Springer Science+Business Media, LLC

H. V. Crock, M.D.
Senior Orthopaedic Surgeon
St. Vincent's Hospital
Melbourne, Australia

H. Yoshizawa, M.D.
Associate Professor of Orthopaedic Surgery
Nagoya-Hoken-Eisei University
Aichi, Japan

Formerly, Leverhulme Research Fellow in Orthopaedics at St. Vincent's
Hospital Melbourne, Australia from Keio University, Tokyo, Japan

Library of Congress Cataloging in Publication Data

Crock, Henry V
The blood supply of the vertebral column and spinal cord in man.

Bibliography: p. 121
Includes index.
1. Spine – Blood-vessels.   2. Spinal cord – Blood-vessels.
I. Yoshizawa, Hidezo, 1937–   II. Title.
QM111.C76   611′.82   76–40960

© 1977 Springer Science+Business Media New York
Originally published by Springer-Verlag New York Inc. in 1977.
Softcover reprint of the hardcover 1st edition 1977

ISBN 978-3-7091-3670-6   ISBN 978-3-7091-3668-3 (eBook)
DOI 10.1007/978-3-7091-3668-3

Production manager: Francis Corless
Production editor: Janet Scott
Designer: Edmée Froment
Printer: R. R. Donnelley & Sons Company

Dedicated to our honored teachers, the late Thomas King
of Melbourne and Professor Torai Iwahara of Tokyo

PREFACE

This book was written in an attempt to fill a serious gap in medical literature. A concise anatomical text, incorporating an atlas of the vascular anatomy of the vertebral column and spinal cord, it has been designed to suit the needs of orthopedic surgeons, neurosurgeons, and neurologists. We also feel that this work will help give direction to further studies on the morphology and physiology of spinal circulation.

The book begins with a discussion of the origins of the arteries which supply the vertebral column. It continues with a description of the various regions of the spine, the cervical, thoracic, and lumbar, and of the arteries which supply the meninges and spinal cord, the veins of the vertebral column, and venous and arterial distribution within the spinal cord.

Anatomic studies of the circulation both in bones and in the spinal cord require the use of techniques which are crude and time consuming. Essential to success is access to fresh postmortem human bodies, so that injections into the vascular systems may be attempted before any dissection has been made in the course of the routine pathologic examination. Permission to pursue these particular anatomic studies cannot be obtained as a matter of routine. Because of the restricted availability of specimens therefore, only random samples can be obtained.

There is a pervading weakness in the literature on the blood supply of the spinal cord. This is that no reference is made to the unusual technical difficulties of completely filling the spinal cord vessels, coupled with the peculiar problems of dissection in this area.

Wide variations are reported in the volumes of substances injected, the injection pressures used and durations of flow of the injected masses. One of the major problems affecting the quality of injection is that of intravascular coagulation. Where death has resulted from violent injury, not only does disseminated intravascular coagulation obstruct vessel filling, but widespread vascular damage allows leakage and loss of injection pressure, so that complete vascular filling is impossible. Refrigeration of the body after death may also adversely affect the quality of injection, especially of smaller vessels.

The thoracic spine is covered by a smaller bulk of muscle than the cervical or lumbar regions, and it rests directly on the cold metal of the storage trays. The significance of the body temperature relevant to the adverse effects of cold on small vessels in this part of the spine therefore assumes great importance. These vessels are frequently frozen. They are therefore effectively closed to any injected mass and they may be too small to be seen unless they have been distended by injection.

Despite all these difficulties, over which the investigator has no control, it is tacitly implied in most writings that the vascular injections of all specimens referred to in the texts are technically perfect. When it comes to statements relating to the number of vessels or branches of vessels in a given region, we believe that many authors have fallen into the error of counting only injected vessels and of ignoring the fact that there may be important small vessels in the areas which are not injected. Clearly then, statistical analysis, desirable as is its use in biology, cannot be applied even to recently published anatomic studies on the circulation in the bones of the human vertebral column and spinal cord.

Recognizing the limitations imposed by many uncontrollable variables, we present this monograph in the spirit of the famous English surgeon anatomist John Hunter, who believed in the overriding importance of careful observation.

H. V. Crock
H. Yoshizawa

ACKNOWLEDGMENTS

In preparing this work for publication we have received assistance from many people whose help we wish gratefully to acknowledge:

Dr. James McNamara, Senior Government Pathologist, Melbourne.
Mr. H. Pascoe, City Coroner, Melbourne.
Mr. Don Hossack. Dr. A. Davis. Mr. Dermot Morgan. Dr. S. Sihombing. Dr. J. Hughes.
Messrs. L. McMahon, B. O'Shea, P. Skene, J. Skehan and J. McDonald.

We thank especially, our colleague Associate Professor S. Kame of the All India Institute of Medical Sciences, New Delhi, who joins us in the authorship of Chapter 6.

We have received permission to reproduce certain figures and textual matter from the Editors of the *Journal of Bone and Joint Surgery and Clinical Orthopaedics and Related Research*, for which we thank them sincerely.

Mr. G. Dommisse, author of *The Arteries and Veins of the Human Spinal Cord from Birth*, and his publishers Churchill Livingstone have given us permission to reproduce certain vessel measurements from that book. We are very grateful for their generous cooperation.

We have received generous financial support from private and public sources and we thank the following donors:

Messrs: G. M. Bedbrook, O.B.E., M.S., M.D., F.R.C.S.
H. Burrowes, E. Dubsky, J. Phillips, A. D. McLean,
P. Norris, H. Williamson.

Miss J. Godfrey, the late Sir Robert Monahan and the late Mrs. Lorna Hannan.

The Australian Orthopaedic Association Research Foundation. The William Buckland Foundation. The Leverhulme Foundation. The National Health and Medical Research Council. The St. Vincent's Hospital School of Medical Research. Nicholas Pty. Ltd., Melbourne.

Professor R. C. Bennett, Professor of Surgery at St. Vincent's Hospital has allowed us to use laboratory facilities in his department and has encouraged us at all times. We are deeply grateful to him.

We thank also Dr. W. M. C. Keane, Medical Superintendent of St. Vincent's Hospital, Melbourne. The Sisters of Charity who at all times have helped us to pursue this work. Dr. Terry Rush who kindly translated Kadyi's paper from the German.

The work involved in the preparation of even a small book such as this requires a great deal of cooperation and understanding from the authors' families; we thank our wives and children for their encouragement.

Finally we thank our patient secretary and manuscript typist Mrs. D. Litchfield for her devotion to duty.

CONTENTS

Preface   vii
Acknowledgments   ix

1

Origins of arteries supplying the vertebral column

Origins of arteries supplying the vertebral column   1

THE CERVICAL REGION   2
THE THORACIC REGION   3
THE LUMBAR REGION   4

2

Origins of arteries supplying the meninges and spinal cord

Origins of arteries supplying the meninges and spinal cord   23

THE MENINGES   24
THE ARTERIES ON THE SURFACE OF THE SPINAL CORD   24
THE SEGMENTAL ARTERIES SUPPLYING THE SPINAL CORD   26

3

Veins of the vertebral column   43

THE EXTERNAL VERTEBRAL VENOUS PLEXUS   43

THE INTERNAL VERTEBRAL VENOUS PLEXUS   44

4

Veins of the spinal cord   55

5

Distribution of arteries within the vertebrae   65

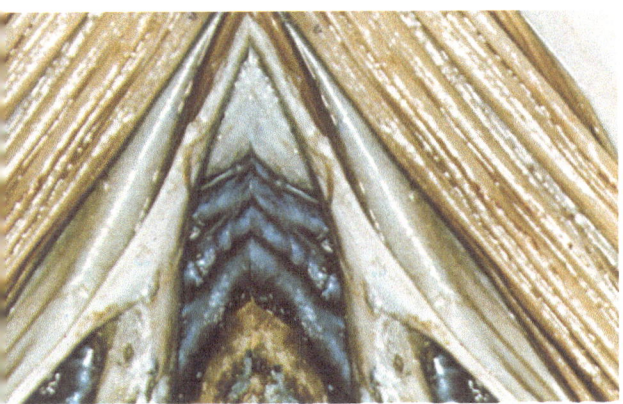

6
Veins of the vertebral body

85

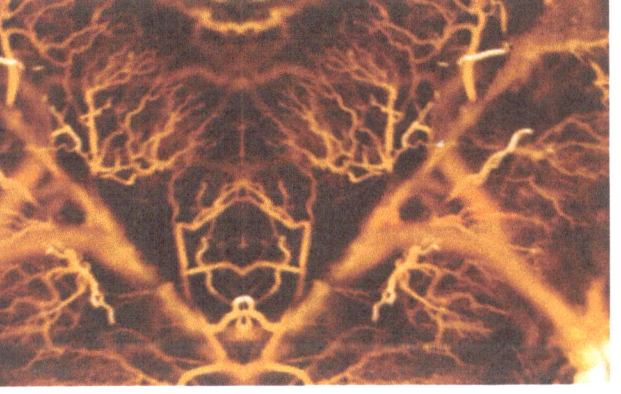

7
Distribution of arteries and veins within the spinal cord

101

THE ARTERIES OF THE SPINAL CORD   102
THE VEINS OF THE SPINAL CORD   103

Notes on Materials and Methods   119
Bibliography   121
Recommended Further Reading   123
Index   125

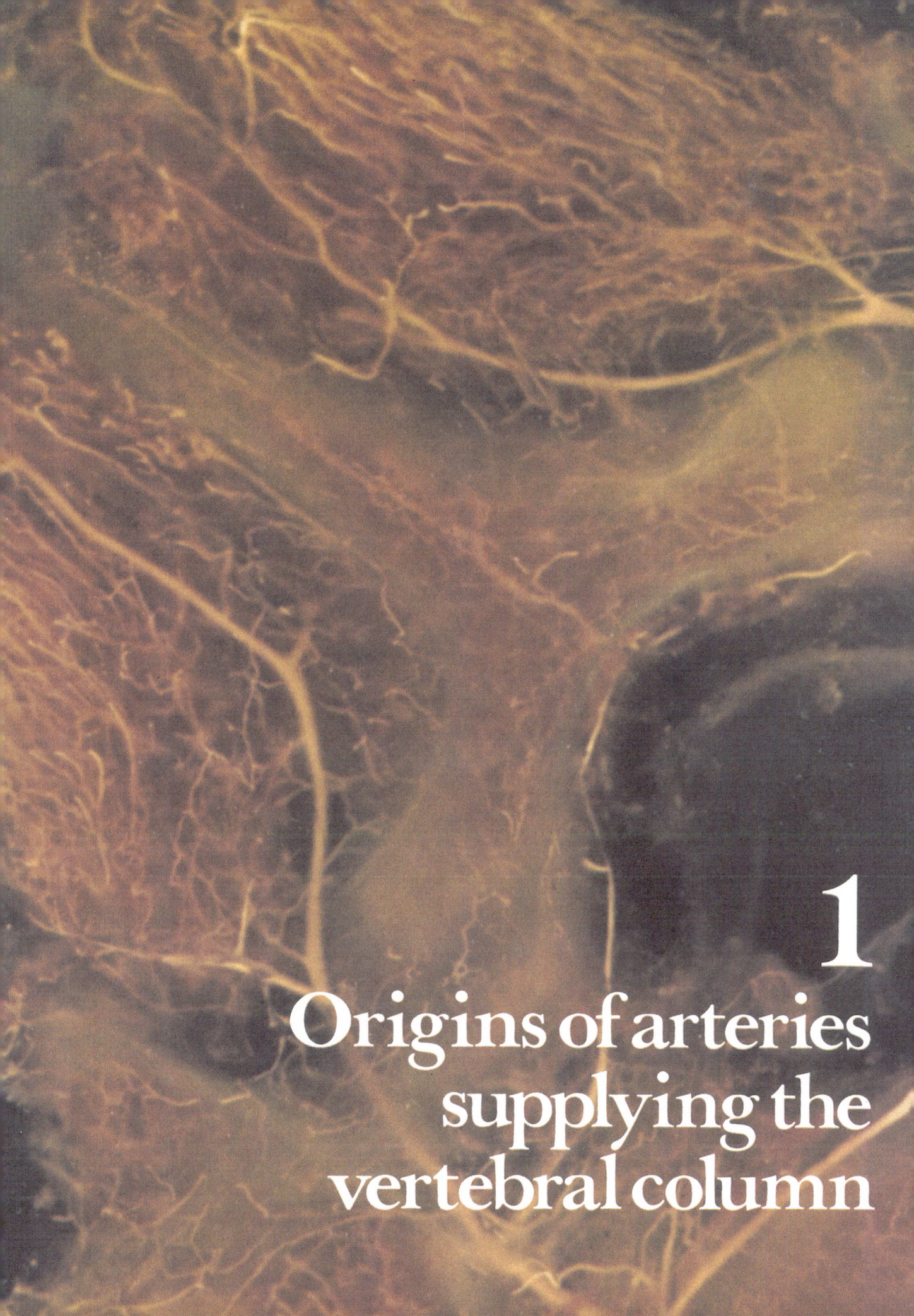

1

Origins of arteries supplying the vertebral column

The cervical vertebrae are supplied largely by the vertebral arteries and by arteries derived from branches of the thyrocervical and costocervical trunks (Figs. 1.1, 1.2).

On the anterolateral surfaces the cervical vertebrae are covered by a complex network of arteries derived from anteriorly directed branches of the vertebral arteries and ascending spinal branches from the thyrocervical trunk.

Each inferior thyroid artery, itself a major branch of the thyrocervical trunk, gives rise to well-defined vertebral branches which run upward and downward forming discrete longitudinal arterial chains along the inner margins of the longus colli muscles. These chains lie on the vertebral column extending from the level of the third or fourth thoracic vertebral bodies below, running roughly parallel until they approach each other near the anterior tubercle of the arch of the atlas above. Inferiorly over the thoracic vertebrae they anastomose with ascending branches from the third posterior intercostal arteries (Fig. 1.3).

From the level of the sixth cervical vertebra upward, branches from the vertebral arteries emerging from beneath the longus colli muscles also contribute to the formation of these longitudinal arterial chains. Beneath the longus colli muscles these anteriorly directed branches from the vertebral arteries are applied to the sides and front of the cervical vertebrae, coursing around the middle of each body from C6 to C2. These branches are analogous to the segments of the posterior intercostal arteries below T5, and to the corresponding segments of the lumbar arteries, both in respect to their courses around the vertebral bodies and to the intrinsic arteries which they supply to them.

Adjacent anteriorly directed cervical vertebral body branches from the vertebral arteries are connected by fine ascending and descending vertical branches which form an open plexus deep to the longus colli muscles on the antero-lateral surfaces of the vertebral bodies.

A regular ladder pattern of arteries is completed on the anterior surface of the cervical and upper thoracic spine by transverse anastomoses which cross between the vertical chains in the middle of each vertebral body.

On the posterior surfaces of the cervical vertebral bodies a regular arcuate system of arteries is formed based on anterior spinal canal branches of the vertebral arteries on each side. These arches sweep around each pedicle, forming a continuous series of linked arcades, with transverse anastomoses linking right- and left-sided arches in the middle of each vertebral body under cover of the posterior longitudinal ligament (Fig. 1.4).

The *anterior spinal canal branches* arise from single fine branches of the vertebral arteries just outside the cervical intervertebral foramina. Similar branches are seen also in the thoracic and lumbar areas to be described below. These arteries are closely related to the pedicles of the cervical vertebrae from C2 downward. They bifurcate into ascending and descending branches on entering the spinal canal, their branches forming the arcade system described above. In the lower cervical region one or more of the anterior spinal canal arteries of the vertebral bodies may be replaced by branches from the deep cervical artery.

The vertebral arteries also supply branches to the cervical nerves and spinal cord, with its covering membranes. These *nervous system branches* are described in Chapter 2.

The *posterior spinal canal arteries* derived from the vertebral artery systems supply the posterior spinal elements, the outer surfaces of which are also covered by plexuses based on the ascending and deep cervical arteries and by descending branches of the occipital arteries. Further details of laminar arteries are described in the section on lumbar vertebrae below.

THE THORACIC REGION

The upper two thoracic vertebrae are supplied by branches of the superior intercostal arteries and anastomoses with descending spinal branches of the thyrocervical trunks. The superior intercostal artery descends from its origin at the bifurcation of the costocervical trunk (into deep cervical artery and superior intercostal artery), across the neck of the first rib and medial to the ventral ramus of the first thoracic nerve (Fig. 1.5). In the first intercostal space it gives off the first posterior intercostal artery which is distributed in a manner similar to the distribution of the lower posterior intercostal arteries.

From the level of the third to the twelfth thoracic vertebrae, branches of the posterior intercostal arteries supply these bones. The third, fourth, and fifth posterior intercostal arteries are long vessels which ascend almost vertically from their origins on the posterior wall of the aorta, crossing the front and then the sides of the upper thoracic vertebrae obliquely to reach their respective intercostal spaces (Fig. 1.3). From about the level of the sixth, seventh, or eighth posterior intercostal arteries the vessels course around the middle of the side of each vertebra, as do the lumbar arteries lower in the spine (Fig. 1.6). The fine branches of the third, fourth, fifth, and sixth posterior intercostal arteries destined to supply the vertebral bodies form vertically orientated plexuses on the front and sides of their respective vertebrae and from these branches arise the regular intraosseous patterns described in greater detail in the section on lumbar vertebrae below.

THE LUMBAR REGION

In the lumbar region there are usually four lumbar arteries which arise in pairs from the posterior wall of the abdominal aorta, the orifices for right- and left-sided branches at each level being separated by only a few millimeters (Figs. 1.7, 1.8). These vessels pass laterally on each side remaining closely applied to the center of the front and side of each vertebral body until they reach the intervertebral foramina. In this part of their course the lumbar arteries give off many fine unnamed branches to the retroperitoneal tissues and to the posterior peritoneum. Their major branches are distributed within the psoas muscles and to the lumbar vertebral bodies.

In relation to the vertebral body, each lumbar artery gives off two sets of branches (Fig. 1.9). The first of these are short *centrum branches* which penetrate vascular foramina at regular intervals, subjacent to the lumbar artery. The second are the longer *ascending and descending branches* which form dense networks on the fronts and sides of the vertebral bodies. Their terminal branches penetrate the bone in the area adjacent to each vertebral end-plate, while other branches form fine vertical networks on the surfaces of the anterior longitudinal ligament and discs. The median sacral artery is of smaller caliber than the lumbar arteries. It arises from the back of the aorta just above its bifurcation. Small paired fifth lumbar arteries usually arise from it and these course across the body of the vertebra toward the intervertebral foramina. Branches from the iliolumbar arteries anastomose with the other vessels on the surface of the fifth lumbar vertebra (Fig. 1.10).

At the level of the intervertebral foramina but just outside them, each lumbar artery divides into a series of major branches (Fig. 1.11). On first inspection there is an overwhelming complexity to the patterns of fine arteries in these regions, the intervertebral foramina being the gateways for arteries and veins of the spinal canal and its contents. However, remarkably constant basic patterns of arterial branching can be discerned by analyzing specimens prepared in various ways. Each lumbar artery gives off three main sets of branches: to the body wall, the spinal canal, and the posterior spinal elements.

Anterior (abdominal wall branches)

The course of these arteries does not concern us beyond stating briefly that they lie medial to and then behind the psoas muscles anterior to the lumbar plexus. Neural branches descend with each lumbar nerve. The abdominal wall branches pass laterally across the quadratus lumborum, piercing the posterior aponeurosis of the transversus abdominis, to pass forward between this and the internal oblique muscle.

Intermediate (spinal canal branches)

There are three subdivisions of these branches to be considered:

a. *Anterior Spinal Canal Branches*
Almost immediately on entering the spinal canal this artery bifurcates into an ascending and descending branch, the latter being closely related to the superior border of the pedicle of the lower vertebra at the interspace (Figs. 1.12, 1.13). Each ascending limb crosses the disc in its outer one-third as it passes upward to join the descending branch from the lumbar artery above it, thus forming an arcade system notable for its regularity (see also Fig. 1.4).

b. *Nervous System Branches*
Arteries accompany the nerves both proximally toward the spinal cord and distally. Detailed descriptions of their distribution are found in Chapter 2. Occasionally, variations in the origins and distribution of both the

anterior spinal and the nervous system branches are seen (Figs. 1.13–1.15).

c. *Posterior Spinal Canal Branches*

The caliber of these branches is marginally smaller than the corresponding anterior spinal branches. They too are disposed in an arcuate pattern, though their branches form a more closely woven network on the anterior surface of the laminae and ligamenta flava, from which vessels penetrate the laminae (Fig. 1.16). Usually a well-marked central artery penetrates the base of the spinous process to run backward toward the tip of the process (Fig. 1.11). Unlike the branches of the anterior arcuate system, those stemming from the posterior spinal arcuate arteries run a tortuous course. Fine unnamed branches run in the extradural fat and others pass to join the arterial plexus on the dura mater.

Immediately on entering the spinal canal, a well-defined laminar branch arises from the posterior spinal arterial arcade on each side. This enters the lamina near its junction with the pedicle and bifurcates at once into a shorter ascending and a longer descending limb, both of which course in the center of the cancellous bone of the lamina toward the subchondral bone plates of the superior and inferior apophyseal joints, respectively (Figs. 1.17, 1.18).

Posterior (body wall branches)

These dorsal rami crossing the pars interarticularis, pass backward in contact with the outer surface of the laminae. They enter the sacrospinalis muscles and course medially and backward, being applied closely to the middle of each spinous process, on the surface of which they form an open meshed plexus (Figs. 1.17, 1.19). Again, the density of branches in the paraspinal muscles is such as to obscure the underlying simple design of the vascular arcades. Beautiful arterial arches form around the posterior vertebral (apophyseal) joints, from which tributaries penetrate the outer surfaces of the laminae, and the joints themselves (Fig. 1.19).

As the dorsal rami pass medially toward the spinous processes, they give off vertical branches which ascend and descend in the substance of the paraspinal (sacrospinalis) muscles. From the plexus formed on the outer aspect of the laminae and the spinous processes, many fine arteries penetrate the bone (Fig. 1.20).

The blood supply of the vertebral column
and spinal cord

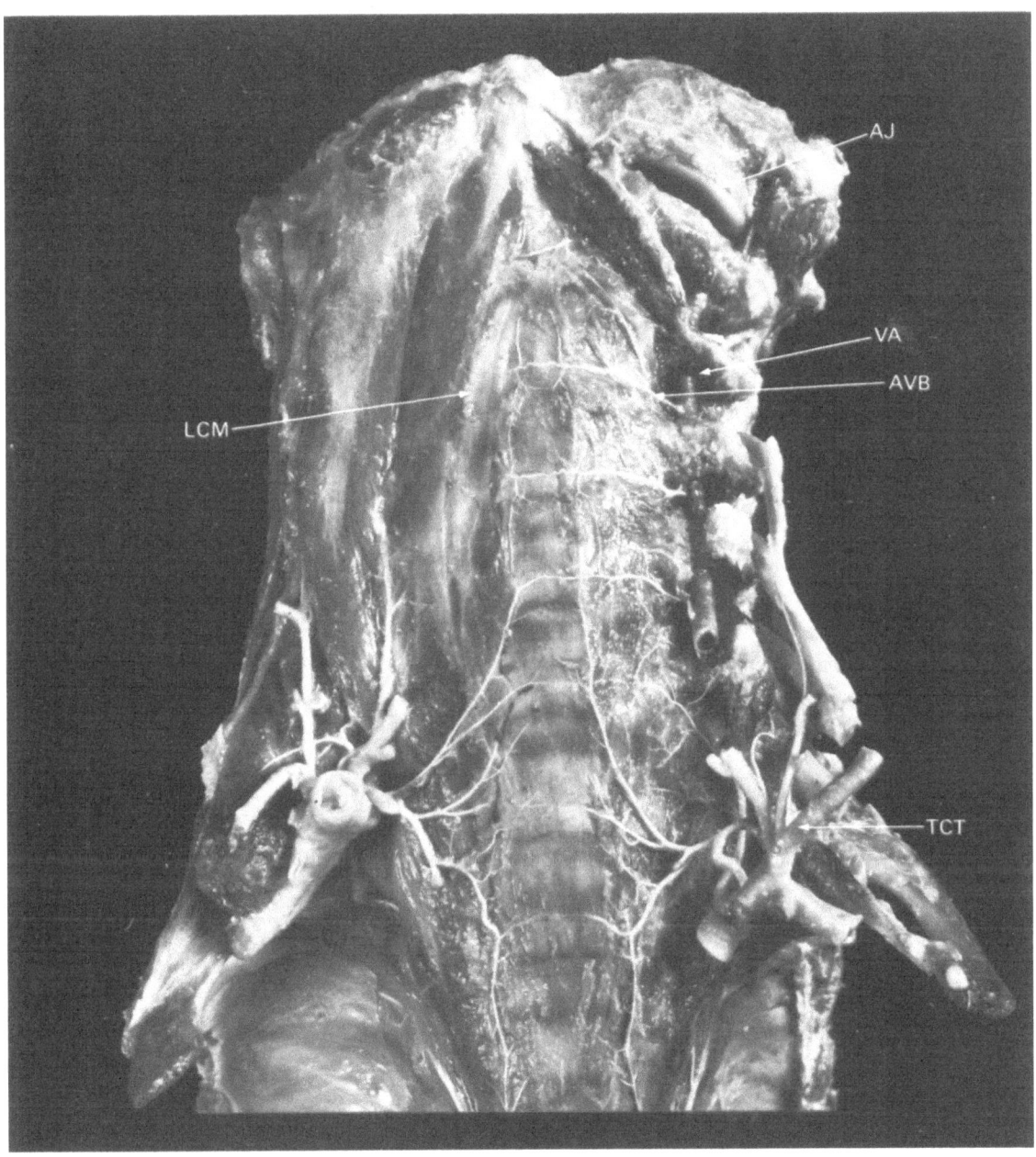

figure 1.1

A photograph of a dissection of the anterior aspect of the cervical and
upper thoracic spine of a female child aged 3¹/₂ years, showing the
origins and course of arteries supplying the anterolateral aspects of
the vertebral bodies. AJ, atlantoaxial joint; LCM, longus colli muscle;
VA, vertebral artery; AVB, anterior vertebral branch of VA (is analogue
of posterior intercostal and lumbar arteries); TCT, thyrocervical trunk.

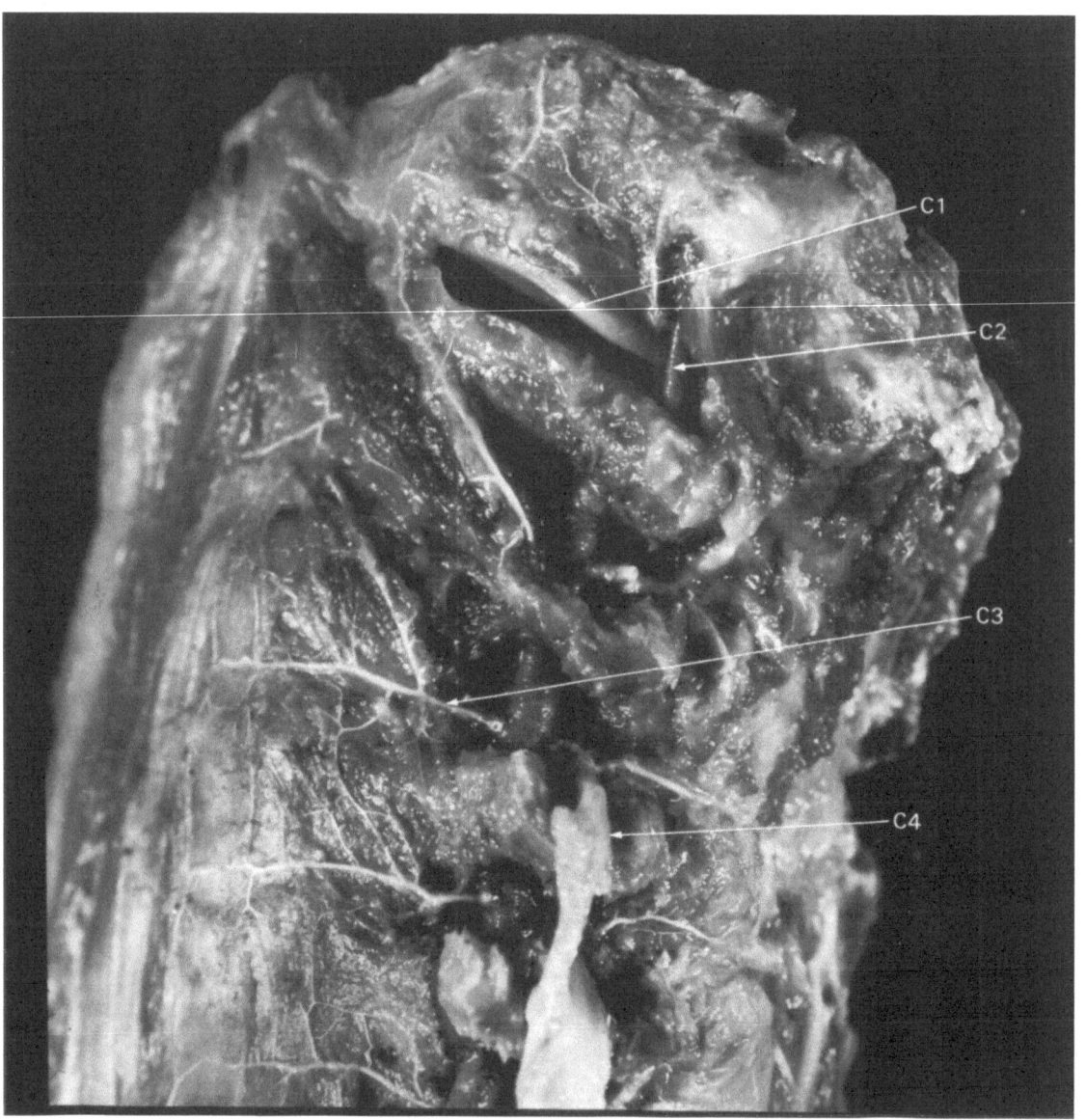

figure 1.2

A detailed view of the specimen illustrated in Figure 1.1, showing the
anterolateral vertebral body branches arising from the vertebral artery
at the level of C1, C2, C3, and C4 vertebral bodies. C1, atlantoaxial
joint; C2, vertebral artery; C3, anterior vertebral body branch of the
vertebral artery (analogue of the intercostal and lumbar arteries);
C4, third cervical nerve root.

figure 1.3

A photograph of the specimen
illustrated in Figure 1.1, showing the
left anterolateral aspect of the
cervicothoracic junction. The long
upward oblique course of the third
posterior intercostal arteries (3rd
LPICA and 3rd RPICA) on both sides
can be seen from their origin at the
juncture of the arch and descending
aorta. In the upper part of the
picture, the left superior intercostal
artery (LSICA) can be seen.

figure 1.4

A photograph of a dissection of the
posterior aspect of the cervical and
upper thoracic spine of a male aged
34 years. The posterior aspects of the
vertebral bodies have been exposed
and portions of the posterior
longitudinal ligament have been
removed from a number of vertebral
bodies in the lower part of the
specimen. In the upper part of the
specimen on both sides, the origins
of the anterior spinal canal branches
from the vertebral arteries can be
seen. In the neck these vessels form
the familiar arcuate arterial pattern
on the anterior surface of the spinal
canal which is found along its length.

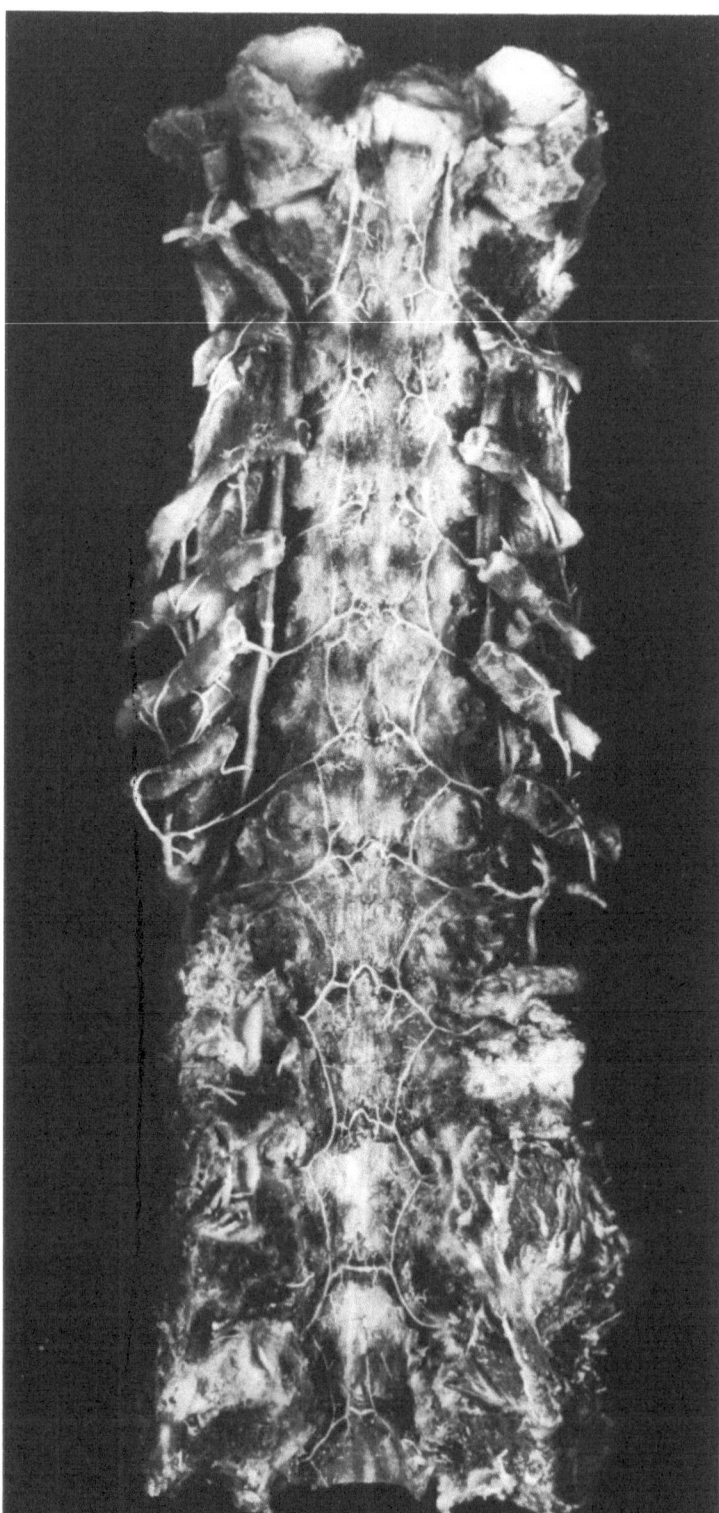

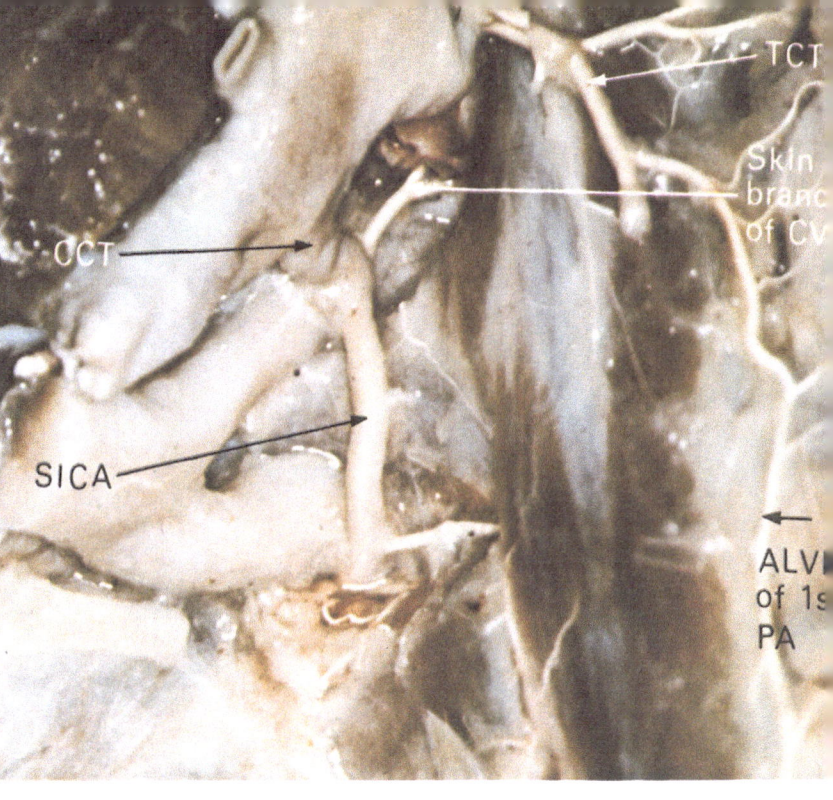

figure 1.5

A detailed photograph of the right
side of the cervicothoracic junction
from the specimen illustrated in
Figure 1.1, showing the costocervical
trunk with the right superior
intercostal artery descending to its
termination in the first intercostal
space. TCT, thyrocervical trunk; CVA,
deep cervical artery; ALVB of 1st PA,
descending spinal branch from the
thyrocervical trunk; CCT,
costocervical trunk; SICA,
superior intercostal artery.

figure 1.6

A lateral view of the specimen
illustrated in Figure 1.1, showing the
right side of the thoracic vertebral
column with the third, fourth, fifth,
sixth, seventh, eighth, and ninth
intercostal arteries dissected.

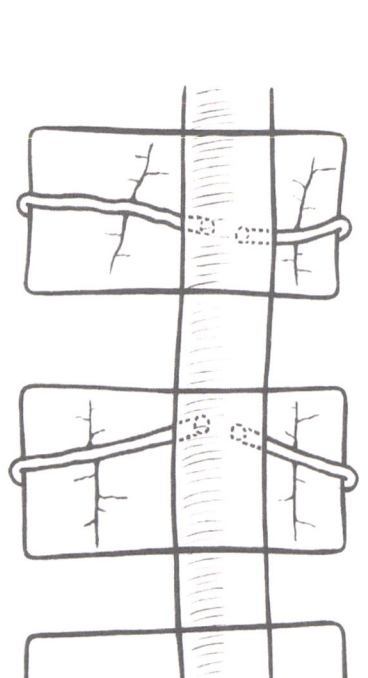

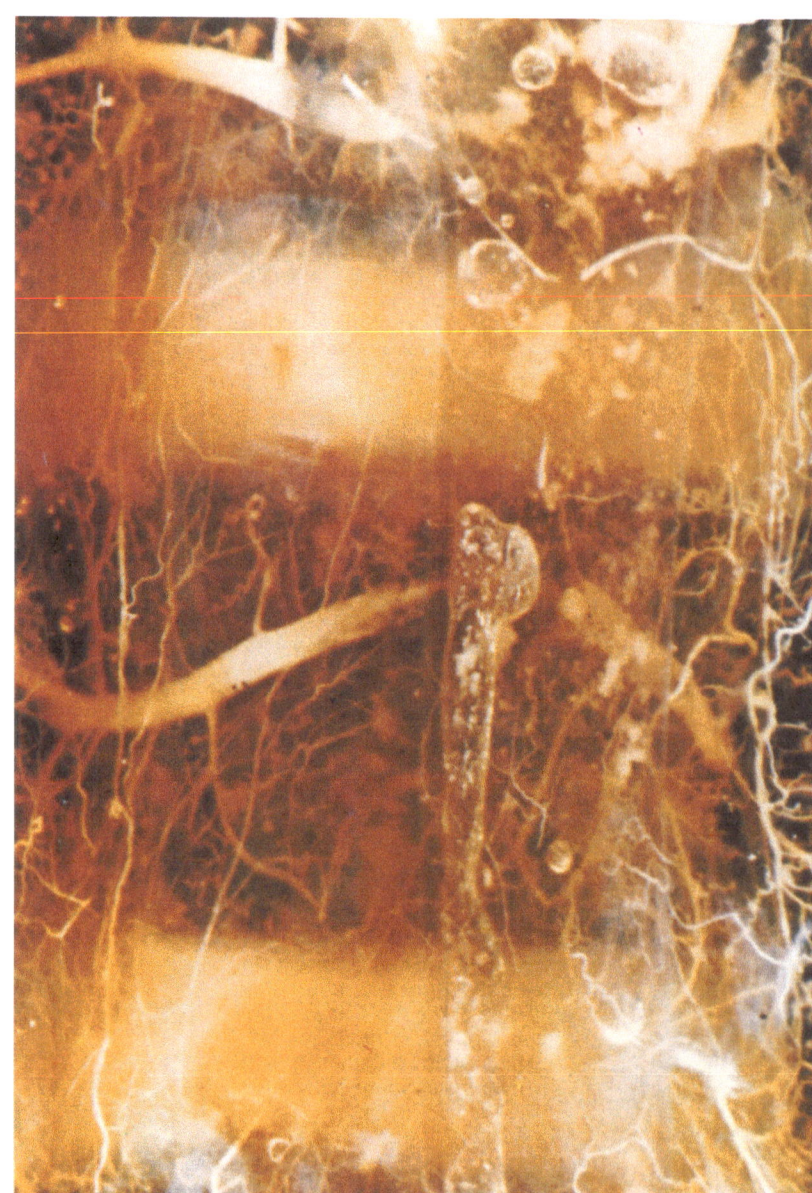

figure 1.7

A detailed photograph of the anterior surface of the lumbar spine of an
adult to show the origin of the lumbar arteries from the posterior
surface of the aorta, with an explanatory diagram alongside.

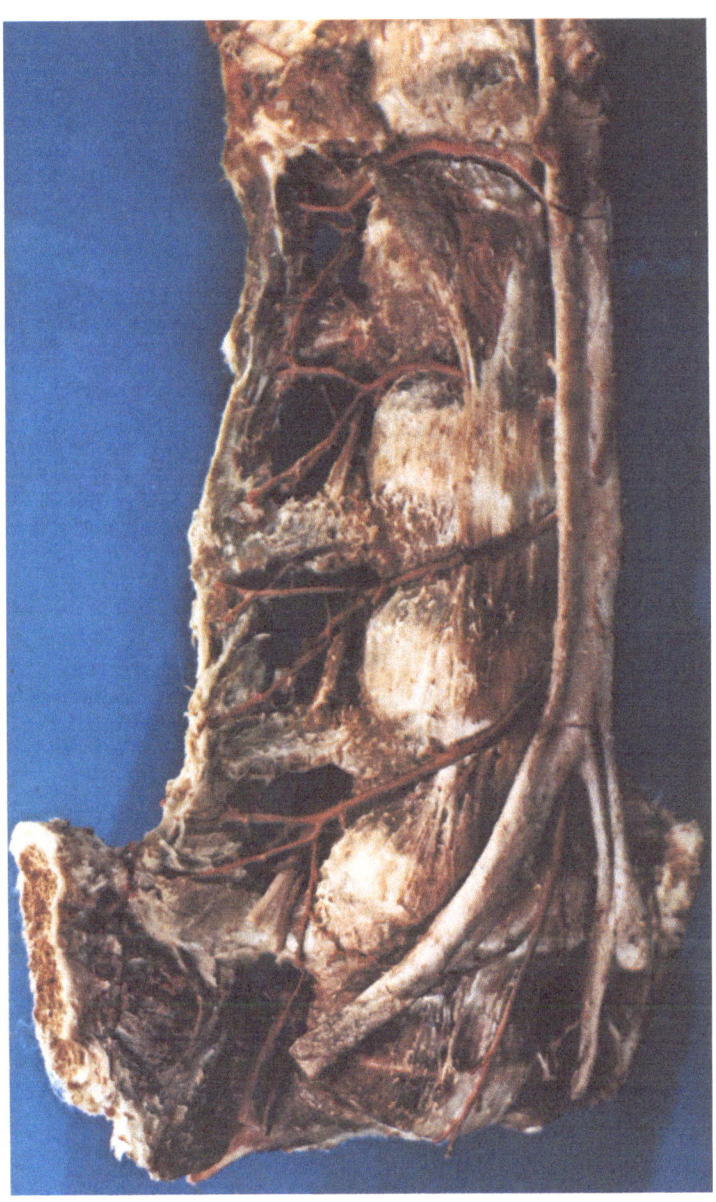

figure 1.8

A photograph of a dissection of the lumbar vertebral column in an
adult viewed from the right side, showing the origins and distribution
of the lumbar arteries from the aorta. The median sacral artery is also
clearly shown. (Dissected by H. Yoshizawa.)

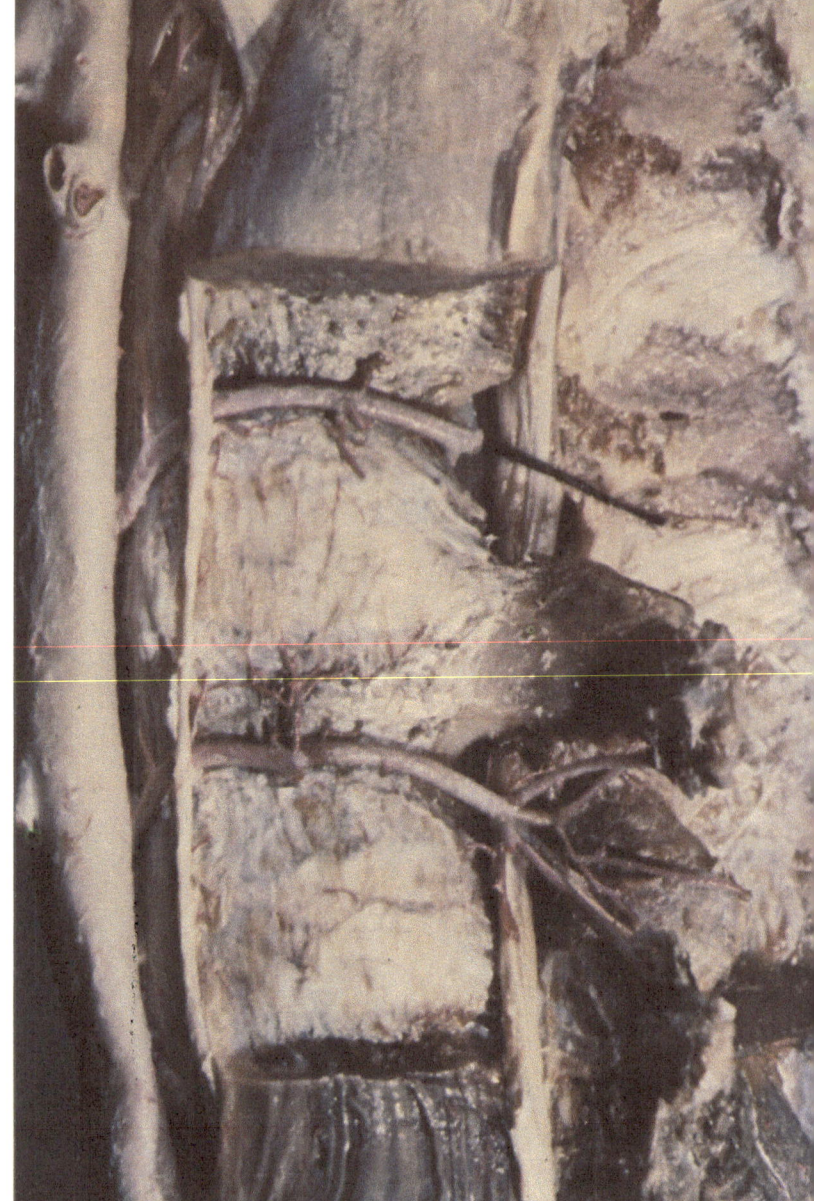

figure 1.9

A photograph of a dissection of the
left side of the lumbar spine of a
child aged 9 years showing the origin
of the centrum branches to the
vertebral body. All other branches
have been cut away except for the
ascending branches arising from the
lower of the two lumbar arteries
shown (arrows). (Reproduced by
courtesy of J. B. Lippincott and
Company from *Clinical Orthopaedics
and Related Research,* No. 115, 1976.
Dissected by H. V. Crock.)

figure 1.10

A detailed photograph of a dissection
at the lumbosacral junction in an
adult, showing the relations of the
third and fourth lumbar arteries
to the sympathetic trunk and the
pattern of branching of the median
sacral artery related to the fifth
lumbar vertebral body. (Dissected
by Dr. S. Sihombing of Indonesia.)

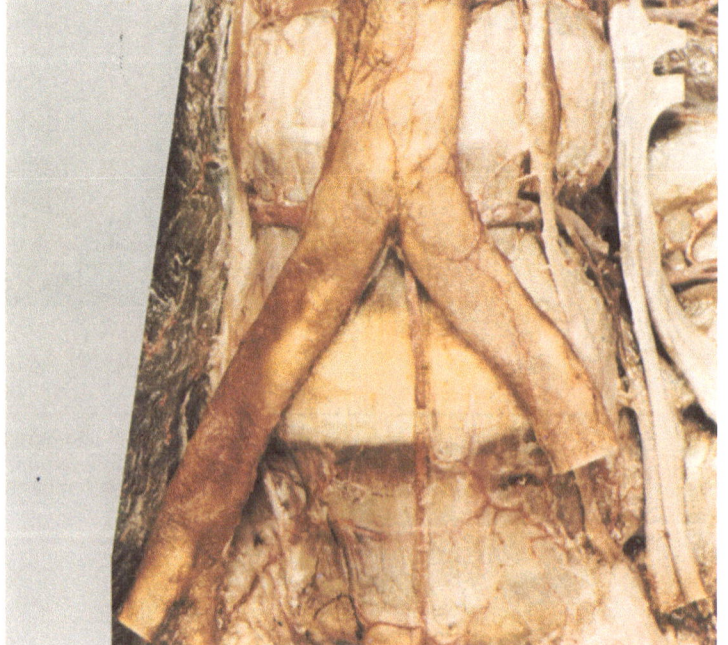

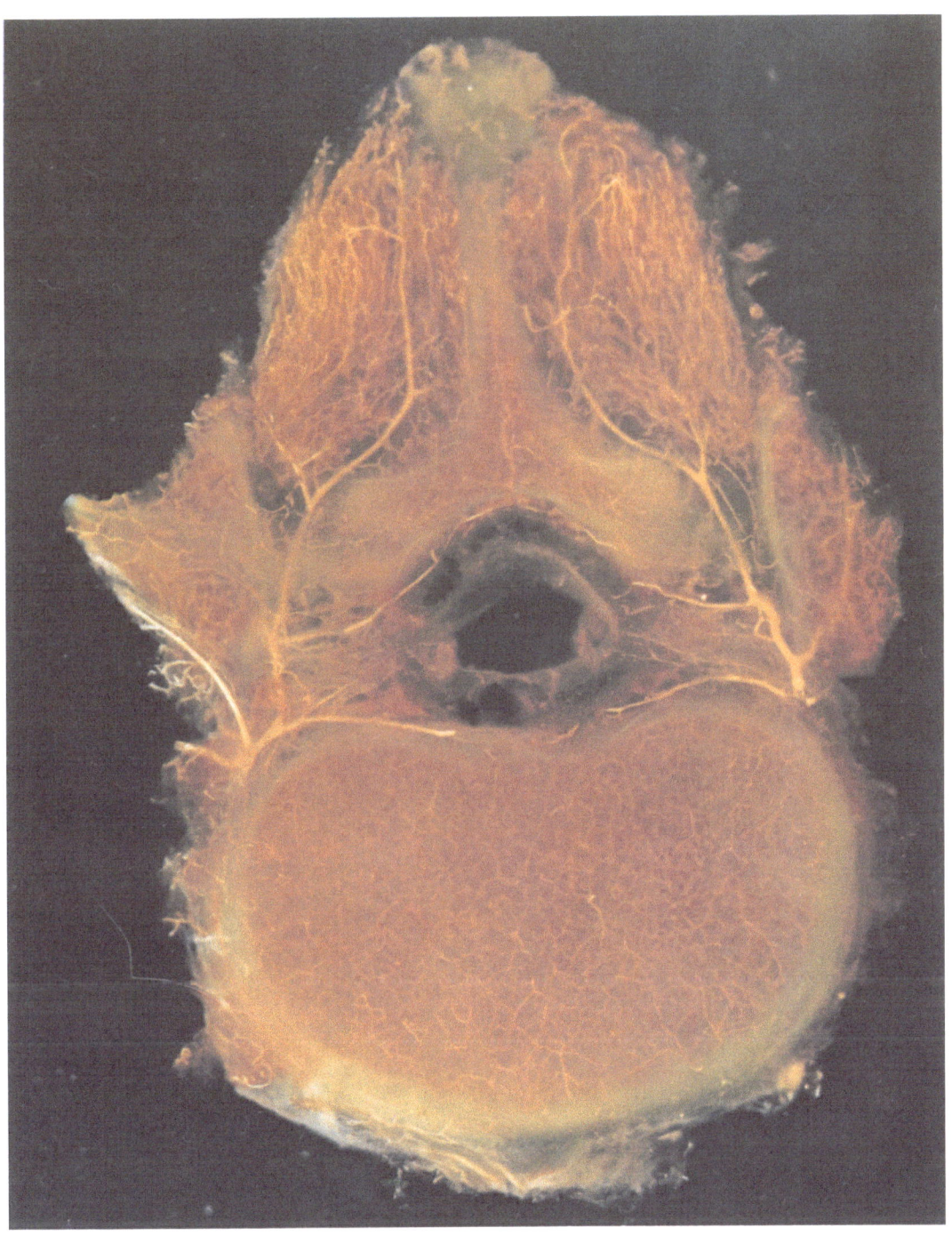

figure 1.11

A photograph of a midlumbar vertebral body cut in transverse section
and viewed from above, showing the main branches of distribution of
the right- and left-sided lumbar arteries from the level of the
intervertebral foramina backward to the tip of the spinous process.
Note: 1, the anterior abdominal wall branches; 2, the intermediate or
spinal canal branches; and 3, posterior to the intervertebral foramina,
the posterior branches in relation to the lamina and spinous process.

figure 1.12

A radiograph of a thin coronal
section through two adjacent
vertebral bodies in a child aged 7
years, to show the distribution of
anterior spinal branches of the
lumbar arteries on the posterior
surface of the vertebral bodies. The
main intraosseous tributaries from
this arcuate system correspond to
the ascending and descending
branches of the abdominal portion
of the lumbar arteries, and in the
centers of the vertebral bodies,
where the right- and left-sided
arcades approach each other,
centrum branches penetrate the
vertebral bodies.

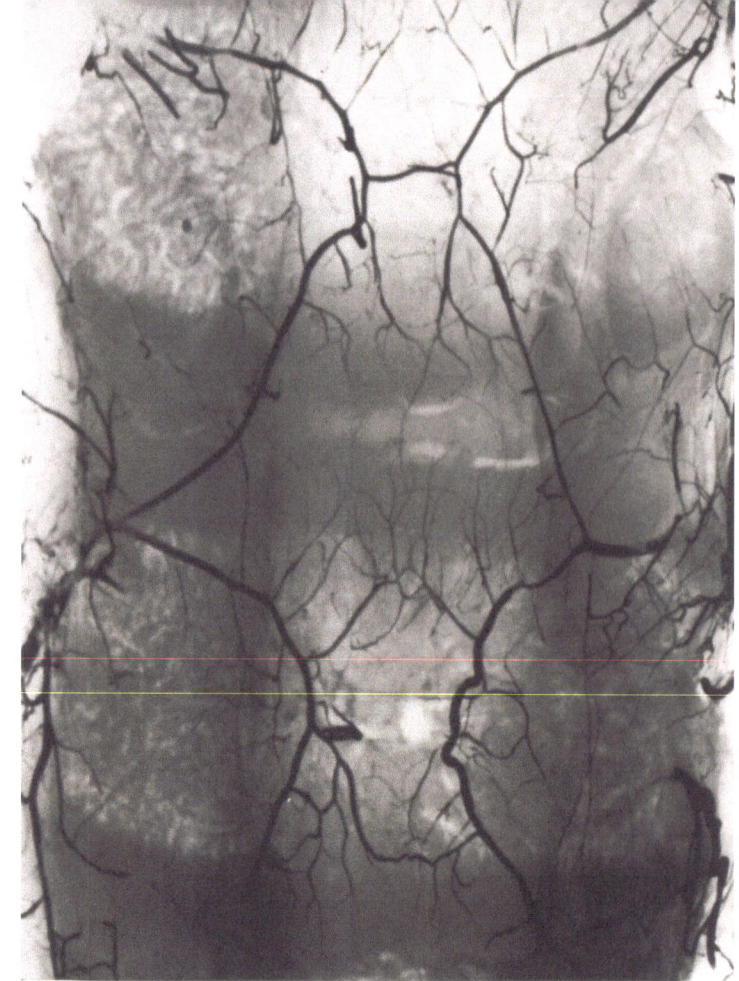

figure 1.13

A detailed photograph to show the usual relationships between the
main stem of the posterior branch of the lumbar artery to the nerve
root at the intervertebral foramen and the nerve root relation to the
anterior spinal canal branches. The dura mater has been removed from
the lower half of the specimen.

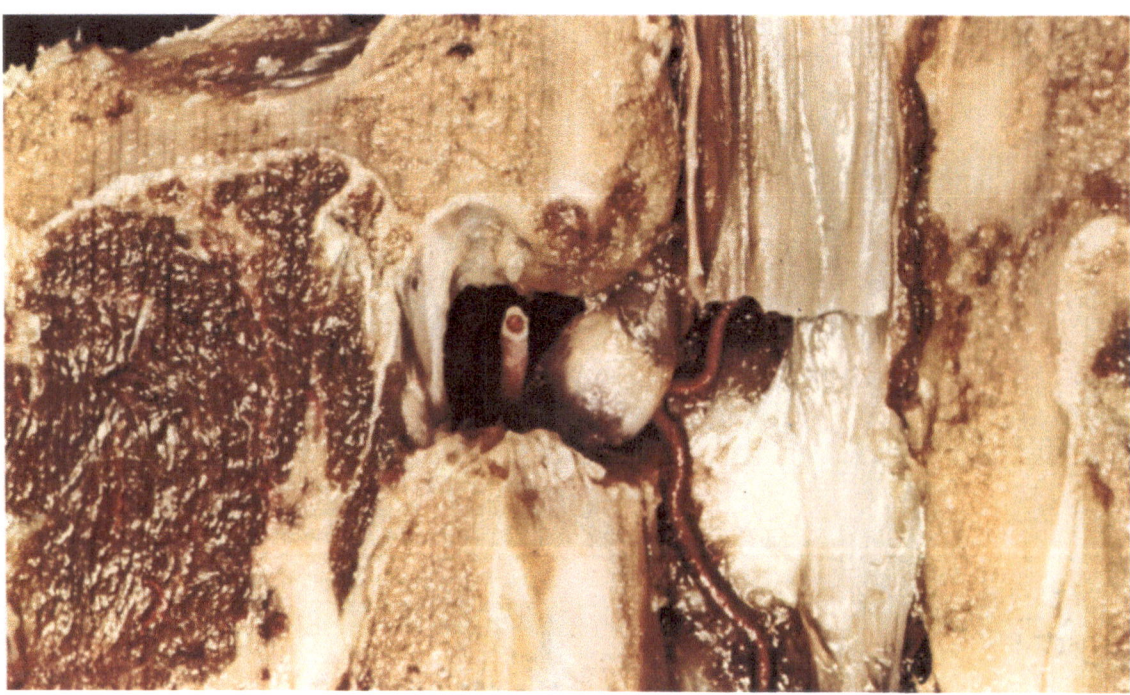

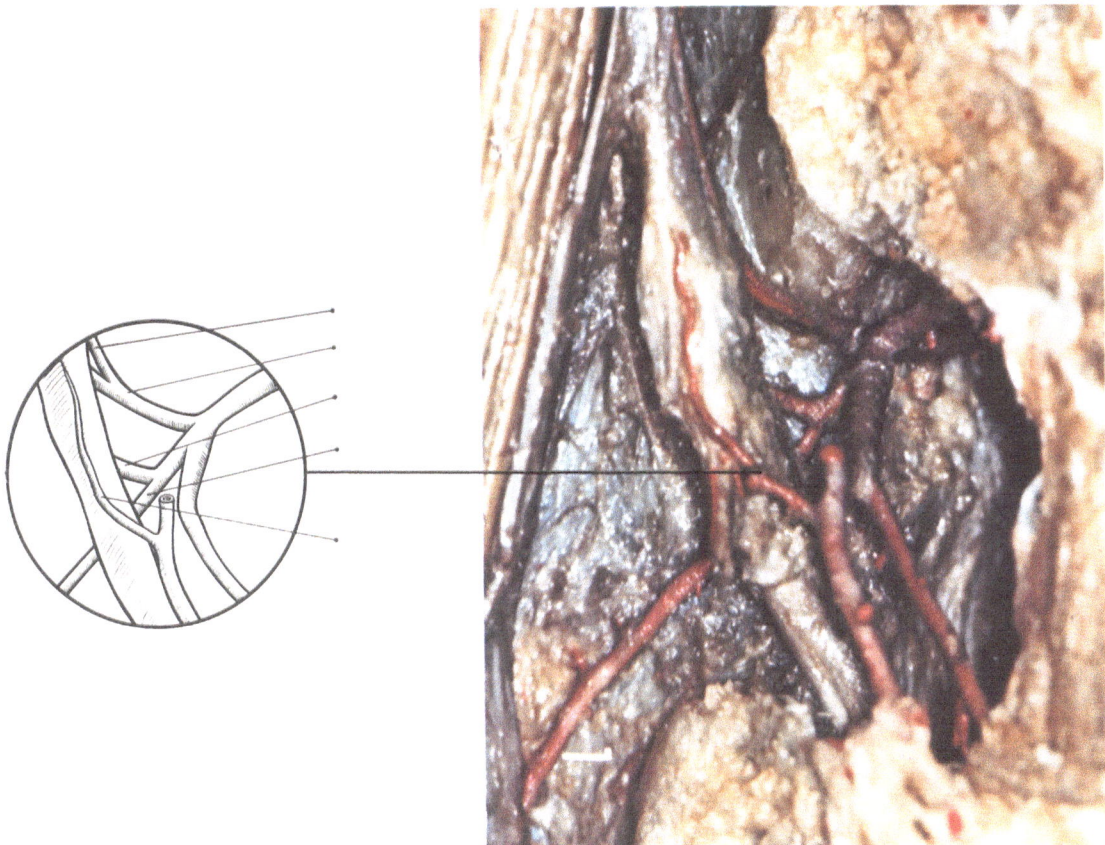

figure 1.14

A detailed photograph to show the anterior spinal canal branches lying
anterior to the emerging lumbar nerve root at the intervertebral
foramen, together with the ascending anterior and posterior nerve root
branches (neural branches) of the lumbar artery, from a female aged
18 years.

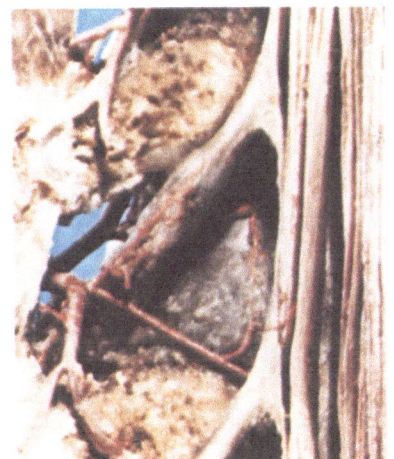

figure 1.15

A detailed photograph from the specimen illustrated in Figure 1.14,
showing a variation in the origin of the anterior spinal canal branch of
the lumbar artery. *Note the inferior arcuate branch lying posterior to
the nerve root.*

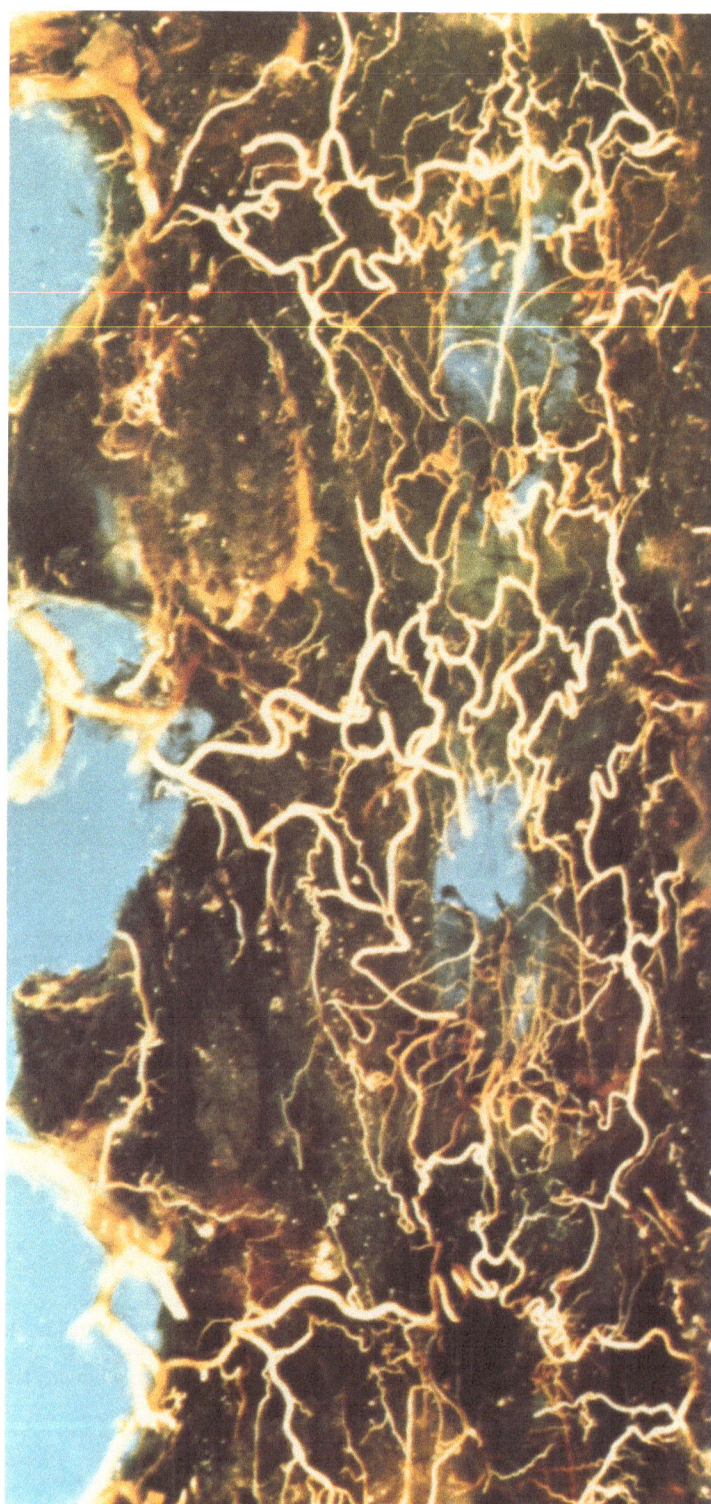

figure 1.16

A photograph of the anterior surface
of the articulated laminae at the
T12/L1 level from a male child aged
12 years. The posterior spinal canal
branches of the lumbar artery system
can be seen entering the canal on
the left side of the specimen as
single branches which bifurcate to
form an ascending and descending
arcuate pattern, corresponding with
the arcuate system formed on the
anterior aspect of the spinal canal.

figure 1.17

A detailed photograph from the
midlumbar spine of an adult. Arterial
injection, Spalteholz cleared
specimen, viewed from outside the
intervertebral foramen, to show the
main stem of the posterior branches
of two adjacent lumbar arteries
as they cross the pars
interarticularis of their respective
lamina, giving off arcuate branches
around the facet joints.

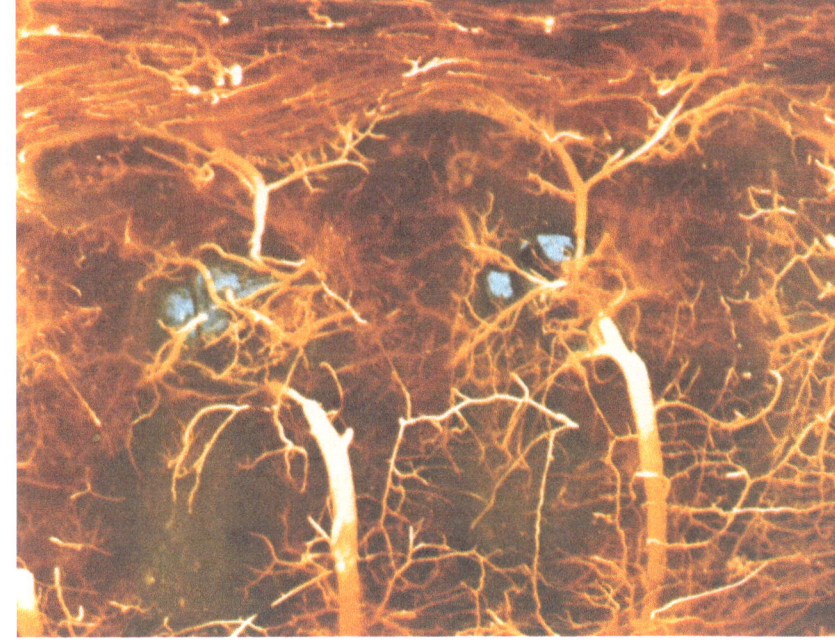

figure 1.18

A detailed photograph of the specimen illustrated in Figure 1.17, viewed
from within the spinal canal, with an explanatory diagram alongside.
Laminal branches are seen in the center of the specimen. The main
vessel enters the lamina near its junction with the pedicle, bifurcating
at once into a short ascending and a long descending limb.

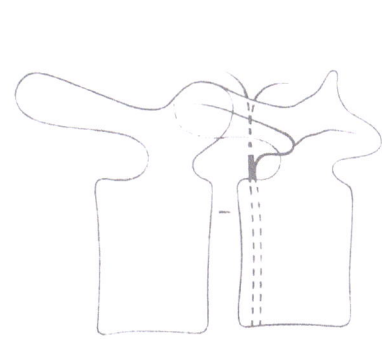

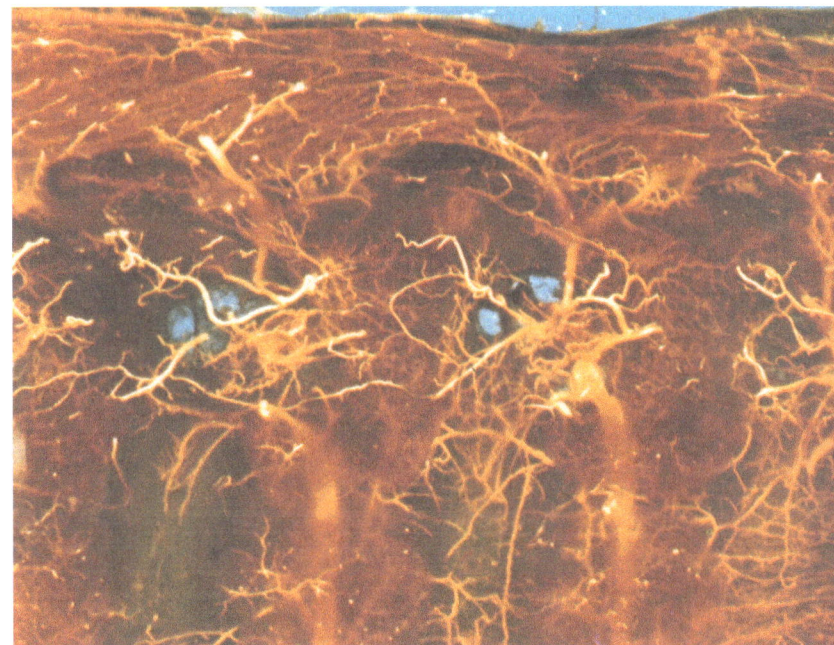

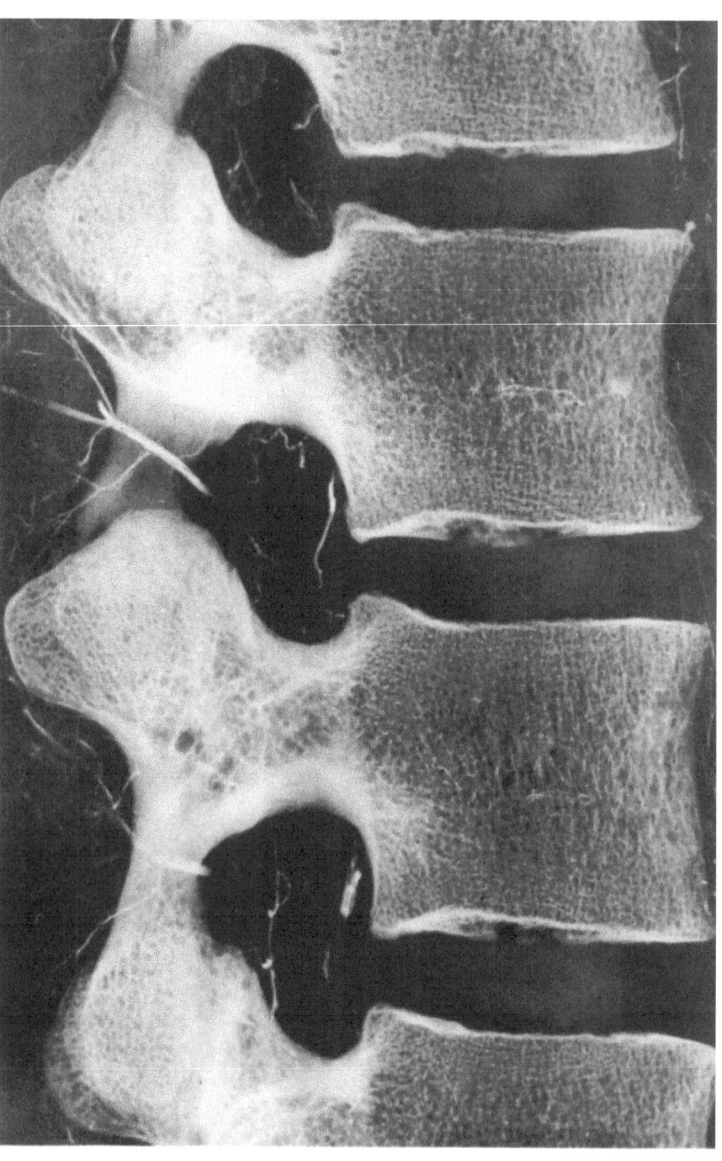

figure 1.19

A radiograph of a thin sagittal section of the lumbar spine, showing
the relations of the posterior branches of two lumbar arteries as they
course backward behind the intervertebral foramina. These arteries
are constant lateral relations of the pars interarticularis of each lamina.
In the lowest intervertebral foramen shown, the anterior and posterior
spinal canal branches of the lumbar artery can be seen, separated by a
clear band, this area being occupied by the nerve root at that level.
Note also the branches which encircle the facet joint system to supply
it. (Reproduced by courtesy of J. B. Lippincott and Company from
Clinical Orthopaedics and Related Research, No. 115, 1976.)

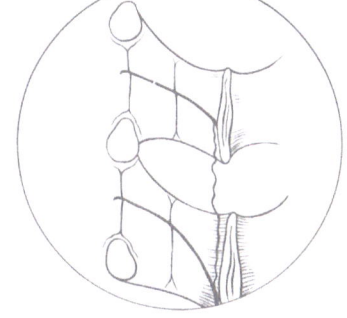

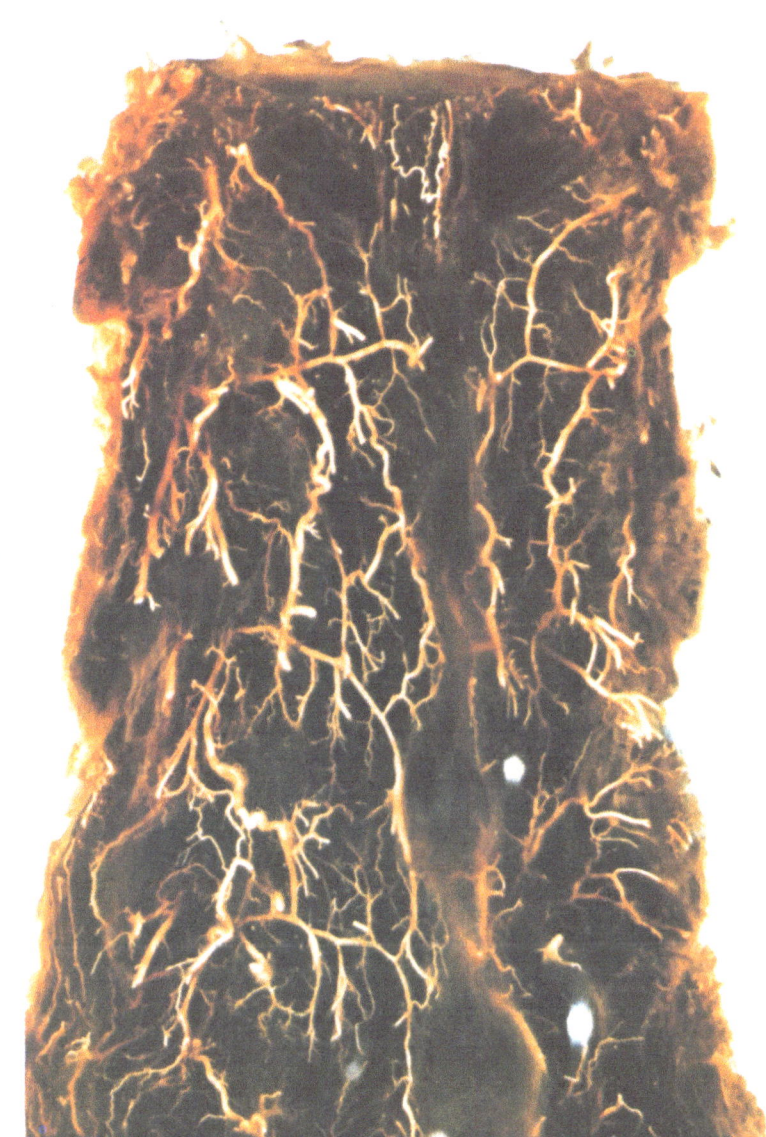

figure 1.20

A photograph of the posterior aspect of the upper lumbar spine of an
adult prepared by the Spalteholz method, following injection of the
lumbar arteries. The specimen is slightly tilted to the right side to
provide a clear view of the posterior spinal branches of three adjacent
lumbar arteries. An explanatory diagram is shown alongside. This
photograph can be interpreted more easily by referring also to Figures
1.11 and 1.19. Note the ascending and descending vertical muscular
branches and the arcuate systems formed around the facet joints.

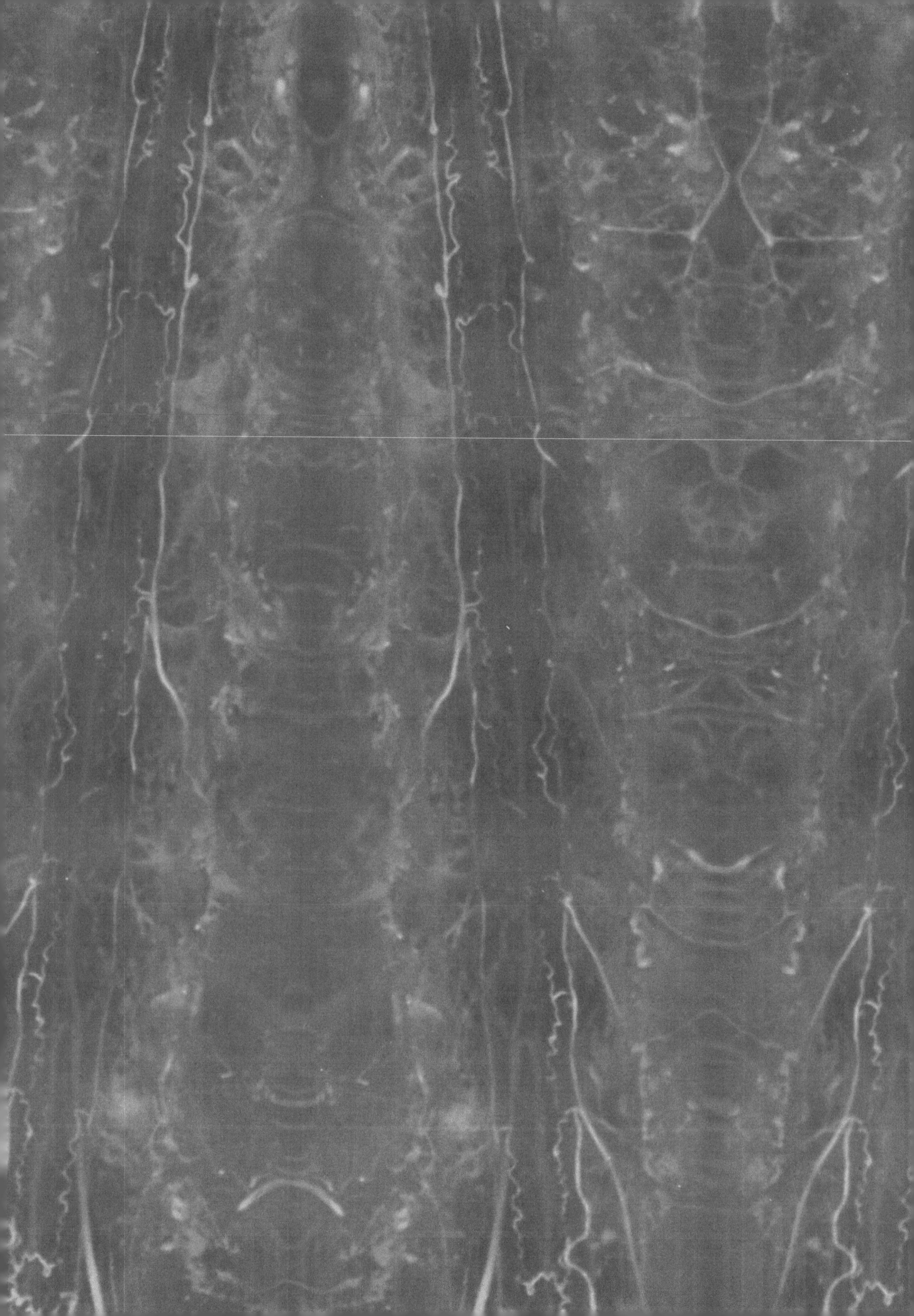

2
Origins of arteries supplying the meninges and spinal cord

The vessels on the spinal meninges are very fine and usually not easily seen in life. They are, nonetheless, arranged in regular patterns reminiscent of the periosteal circulation on long bones (Crock, 1967). In the cervical region (Fig. 2.1), fine branches from the ascending and deep cervical arteries run along the posterosuperior borders of the cervical nerve roots. Just inside the spinal canal, these branches bifurcate to form a longitudinally orientated channel in the epidural space. From this channel, midway between adjacent nerve roots, main stem branches pass on to the side of the dural sac, running transversely backward toward the midline where they anastomose with corresponding arteries from the other side. These unnamed segmental arteries are joined into an open plexus on the side and posterior surfaces of the dural sac by an irregular number of longitudinal branches (Fig. 2.2). A similar pattern is found on the anterior surface of the dural sac, based on meningeal branches of the nerve root arteries.

Over the remainder of the dural sac the meningeal plexus described above is made up of branches from the nerve root arteries, with some contributions from anterior and posterior intraspinal branches of the segmental intercostal, lumbar, and sacral arteries.

THE ARTERIES ON THE SURFACE OF THE SPINAL CORD

Three main arterial channels are found on the surface of the spinal cord. Traditional anatomic terminology defines these respectively as a single anterior spinal artery and paired posterior spinal arteries. These are joined by an open meshed pial plexus around the circumference of the cord (Figs. 2.3, 2.4).

There is general agreement in the literature on the origins of these arterial systems at their cranial end. The anterior spinal artery is formed in the region of the lower end of the olives of the medulla oblongata by the confluence of branches from the right and left vertebral arteries. Likewise, the posterior spinal arteries may arise either from branches of the vertebral or posterior inferior cerebellar arteries over the medulla oblongata.

Controversy surrounds the questions of the site, number, and nomenclature of contributing branches to these three main arterial channels throughout the length of the spinal cord.

Gillilan (1958) suggested certain alterations to standard terminology based on the assumption that not all nerve root arteries reached the spinal cord. Dommisse (1975) has published a terminology for these three major vessels and their tributaries which is essentially in keeping with Gillilan's views.

We favor the adoption of Dommisse's nomenclature for the three major arterial channels on the surface of the spinal cord:

1. The anterior median longitudinal arterial trunk of the spinal cord (Fig. 2.3).
2. The right and left posterolateral longitudinal arterial trunks of the spinal cord (Fig. 2.4).

While the nomenclature of these major trunks is perhaps only of academic interest, we believe that the role of contributing segmental branches to these channels is fundamentally important. From the time of the original contributions of Adamkiewicz (1881a,b) and Kadyi (1889), workers have repeated their studies only to report variations in the site and number of segmental feeder arteries to the three major surface trunks. More recently Dommisse (1975) has reproduced carefully reconstructed schematic drawings showing the *anterior median longitudinal arterial trunk of the spinal cord* and the *posterolateral longitudinal arterial trunks* in 42 spinal cords. These drawings are simplified in comparison with those published by Kadyi (1889), however, they do not change the earlier concepts on the arterial supply of the cord as described by these earlier workers.

Critical analysis of published radiographs, photographs, and schematic drawings raises doubts about the accuracy of many reported observations. There are factors such as the cause of death, postmortem transport and body storage facilities, and the time interval between death and injection which are beyond the control of investigators in this field. For these and other reasons, the quality of injection can rarely be guaranteed, and so, in many instances dogmatic statements on questions concerning the number of vessels involved in the regional blood supply of the cord cannot be made.

The methods used in our preparation of material have been described in detail on pages 119–120. We have been able to produce a number of specimens by tedious dissection in Spalteholz fluid with the aid of a dissecting microscope, *which lead us to suggest that segmental arteries probably always join the anterior median longitudinal arterial trunk of the spinal cord and also the posterolateral longitudinal arterial trunks. At the same time, we recognize that the caliber of these vessels is variable* (Figs. 2.5–2.8).

This observation may be of great importance as we look to future refinements of *in vivo* microangiography of the spinal cord and to the application of rapidly evolving microsurgical techniques to spinal cord problems.

THE SEGMENTAL ARTERIES SUPPLYING THE SPINAL CORD

In Chapter 1 detailed descriptions are given of the branching of arteries at the level of each intervertebral foramen. Of the three sets of branches which enter the spinal canal, the nervous system branches arise from each segmental artery just medial to the site of origin of the anterior spinal canal branches, but still outside the spinal canal. Anterior and posterior nerve root arteries course upward reaching the superior edge of the adjacent nerve root, running along the dural nerve root sleeve for a short distance before penetrating it (Figs. 2.9–2.11). The artery accompanying the anterior nerve root is of larger caliber than that accompanying the posterior nerve root. Dommisse (1975) has provided measurements, at postmortem, of the outside diameters of these "medullary feeders" in each region of the cord in neonatal and adult specimens.

The vessels of origin for these radicular arteries vary in different regions. The majority arise from the major regional arteries, that is, from the vertebral artery in the neck, with occasional contributions from thyrocervical trunk arteries, from the intercostal vessels in the thoracic zone, and from the lumbar and lateral sacral arteries.

Pending confirmation by other workers of the findings which we publish here *it is our contention that the nomenclature for these branches, which contribute to the extrinsic arterial supply of the spinal cord, should refer simply to anterior and posterior*

POSTMORTEM MEASUREMENTS OF MEDULLARY FEEDERS[a]

Medullary feeders	Average outside diameter (μm)
Anterior cervical	
Neonatal	245
Adult	380
Posterior cervical	180
Anterior thoracic	
Neonatal	282
Adult	400
Posterior thoracic	153
Anterior lumbar and sacral	
Neonatal	245
Adult	405

Posterior lumbar and sacral	Average overall size
Neonatal	150
Adult	225

[a] Reproduced by courtesy of G. F. Dommisse and Churchill Livingstone.

radicular arteries. Terms such as medullary feeders or radiculo-medullary arteries would be superseded. These radicular arteries, which vary considerably in size, join the *anterior median longitudinal arterial trunk* and the *posterolateral longitudinal arterial trunks of the spinal cord*, respectively, at each segmental level of the vertebral column.

When more is known about the control of blood flow in small arteries, then further revision of the nomenclature of spinal arteries such as the artery of Adamkiewicz will be required. There may be many feeder arteries with diameters in neonates not less than 350 μm and in older age groups not less than 450 μm, so that the significance of recognizing a single artery of Adamkiewicz may lose much of its currently imagined importance.

We invite the reader to make a detailed study of the illustrations and legends in this chapter because we believe that they support a different view on the segmental arterial supply of the human spinal cord from any previously published.

The blood supply of the vertebral column
and spinal cord

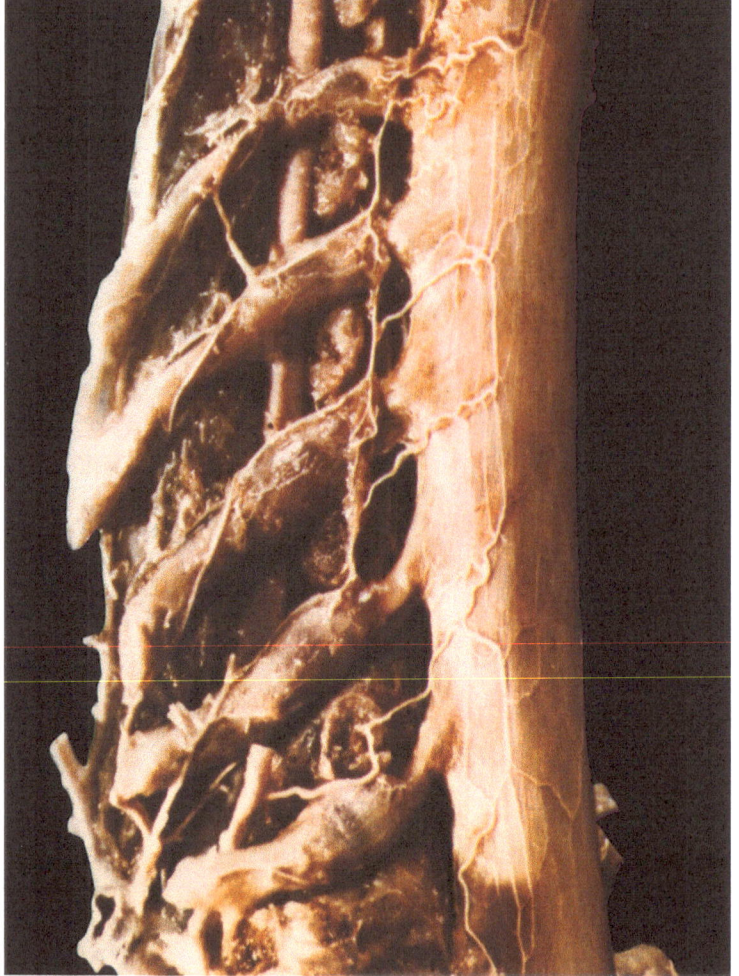

figure 2.1

A photograph of the posterolateral
aspect of the cervical dural sac
showing the origin of the arterial
plexus and its pattern of distribution
on the surface of the dura from a
male aged 34 years.

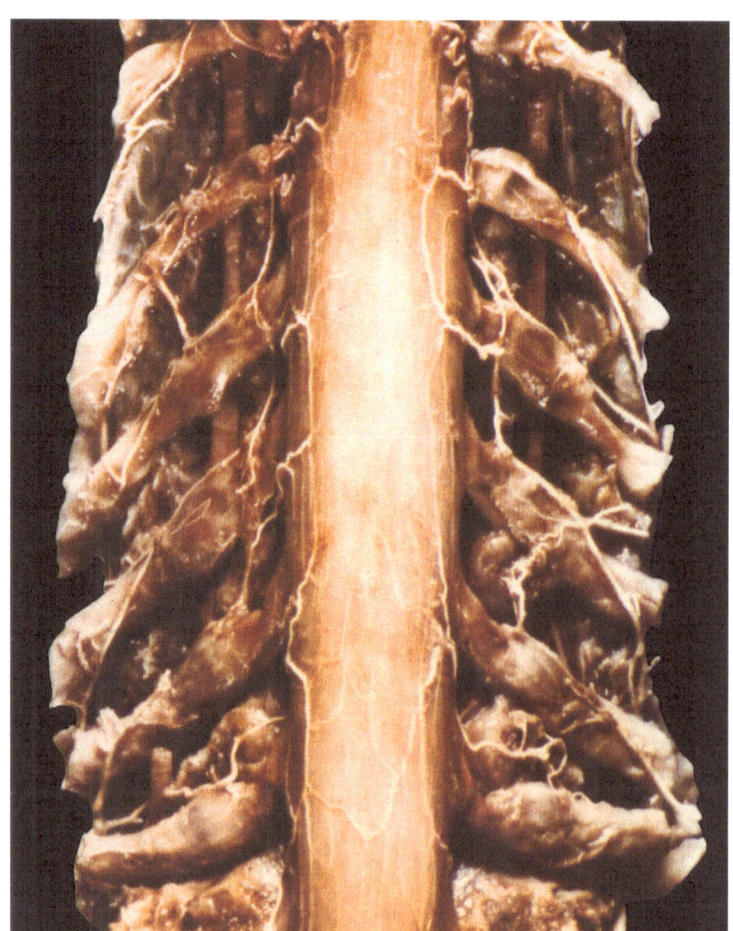

figure 2.2

A photograph from behind of the
specimen illustrated in Figure 2.1,
showing the open arterial plexus on
the posterior surface of the dura.

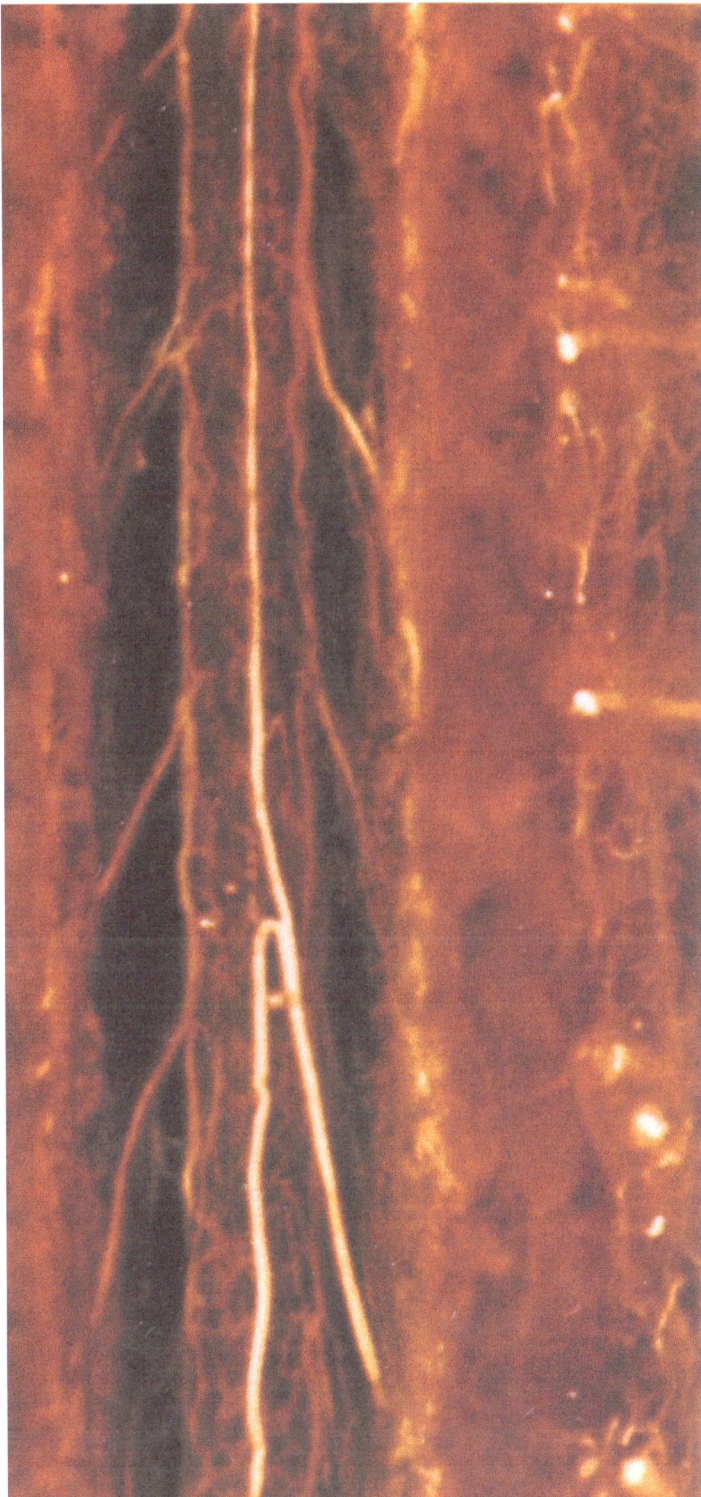

figure 2.3

A photograph showing the *anterior
median longitudinal arterial trunk
of the spinal cord* in a male child
aged 13 years. The length of the
specimen extends from T3 to L2
approximately. In this photograph,
note that only two large
radiculomedullary arteries can be
seen joining the anterior median
longitudinal arterial trunk on the
right side of the photograph. *The
reader must appreciate the
limitations of magnification imposed
on the interpretation of this
observation.*

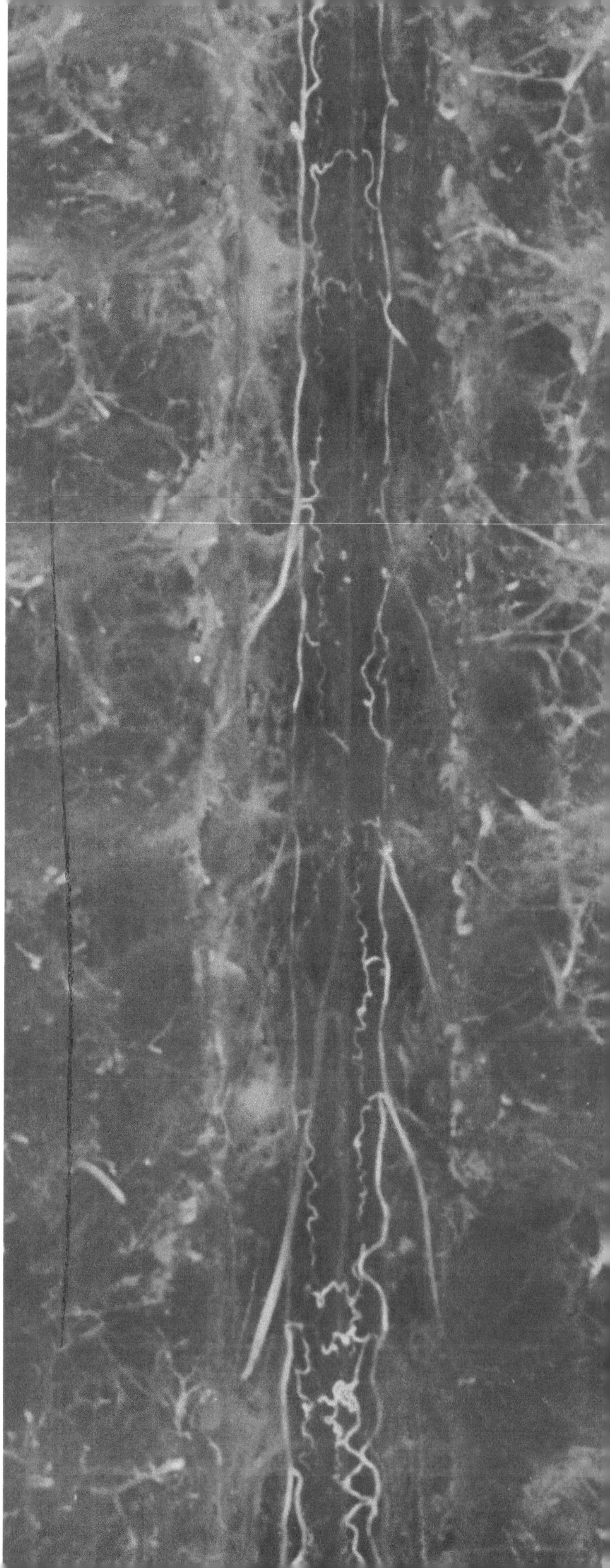

figure 2.4

A photograph of the posterior
surface of the spinal cord illustrated
in Figure 2.3. The arterial system of
the body was injected with
Micropaque. The cord has been
fixed and processed by the
Spalteholz method and dissected
in situ initially by removal of the
posterior arches of the laminae
from T3 to L2 approximately. The
*right and left posterolateral
longitudinal arterial trunks of the
spinal cord* can be seen in focus
with some radicular arteries joining
the channels on either side. *However,
the reader must appreciate that
neither in this figure nor in the
preceding one can any reliable
judgment be made on the numbers
of radicular arteries*
(radiculomedullary arteries) *joining
either the anterior median
longitudinal arterial trunk or the
right or left posterolateral
longitudinal arterial trunks of the
spinal cord.*

figure 2.5

A photograph of the anterior surface
of the cervical spinal cord from a
female aged 20 years. The anterior
median longitudinal arterial trunk
of the spinal cord can be seen to be
duplicated in a number of areas,
while in the lower cervical segment
it is a single large vessel.

This arterial injection of the spinal
cord was prepared in the manner
described in the text, processed, and
dissected in Spalteholz fluid. *In its
unretouched state it provides
evidence supporting the claim that
radicular artery contributions to
the anterior median longitudinal
arterial trunk of the spinal cord are
segmental. The size of individual
radicular arteries varies considerably.*

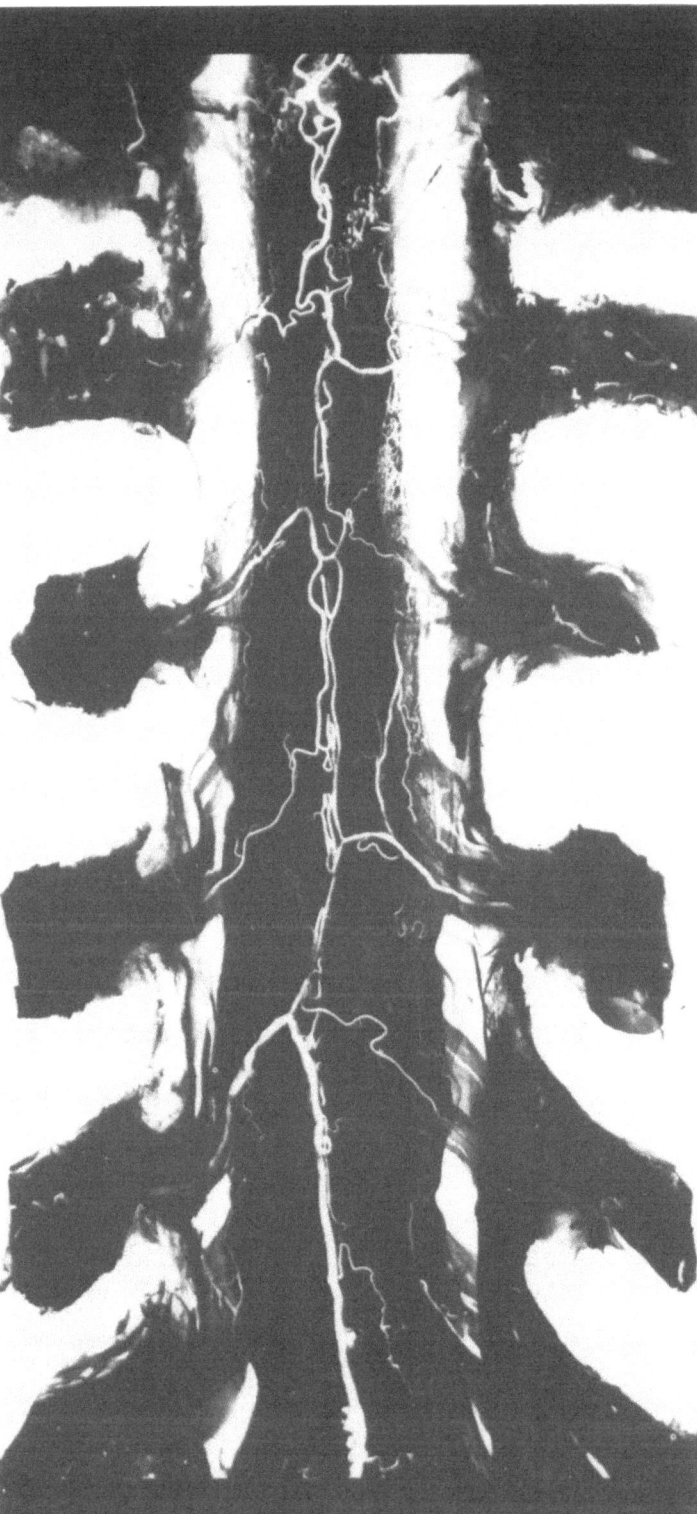

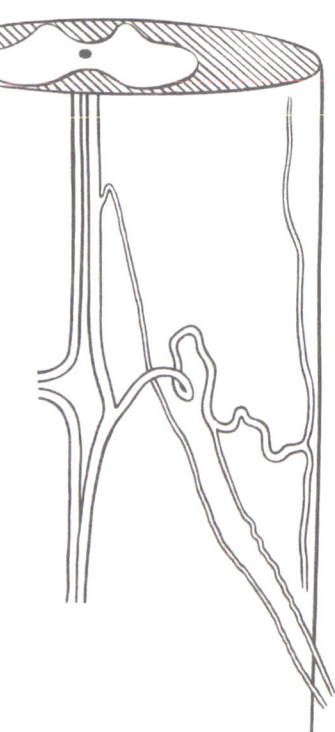

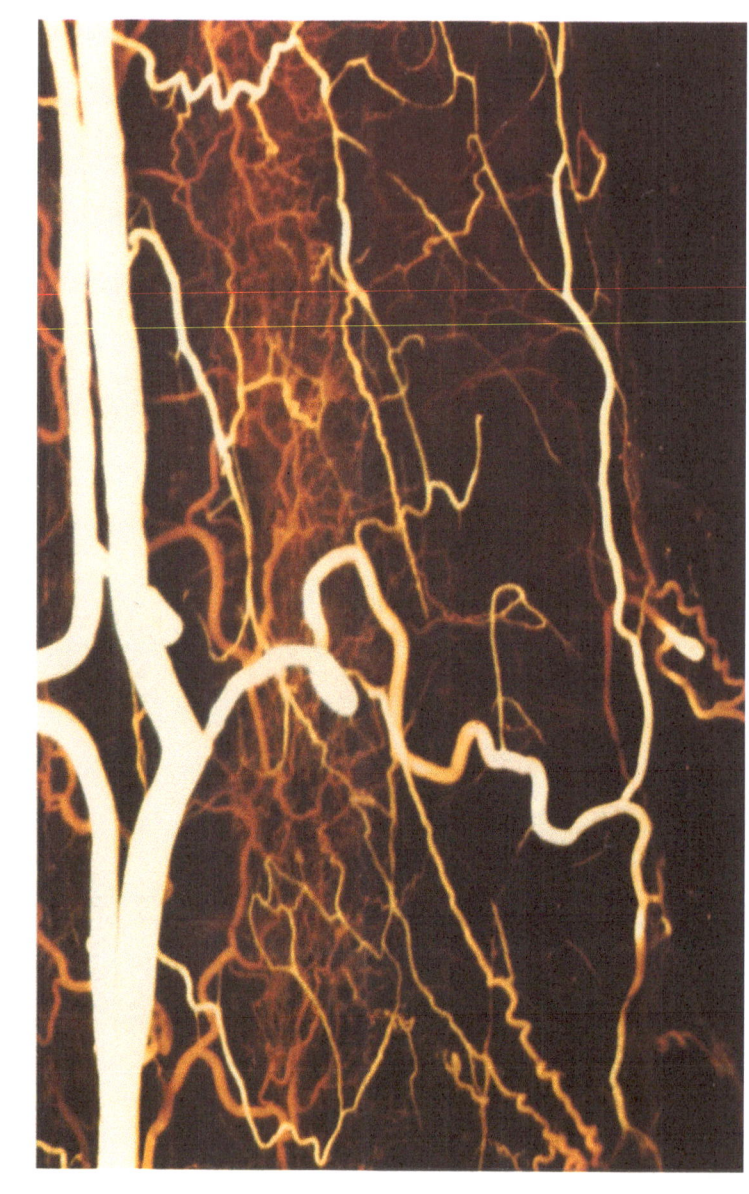

figure 2.6

A detailed view of the *anterior median longitudinal arterial trunk* at the
level of the sixth cervical nerve root from a male aged 45 years,
showing one of the classic diamond patterned reduplications that may
occur in this channel. Spalteholz preparation. A large lateral pial plexus
artery can be seen and two nerve root arteries are visible as outlined in
the accompanying schema. The more laterally placed nerve root artery
joins the pial plexus branch. The longer, medially placed radicular
artery joins the anterior median longitudinal arterial trunk of the spinal
cord. Note also the central arteries and their branching patterns in the
depths of the cord.

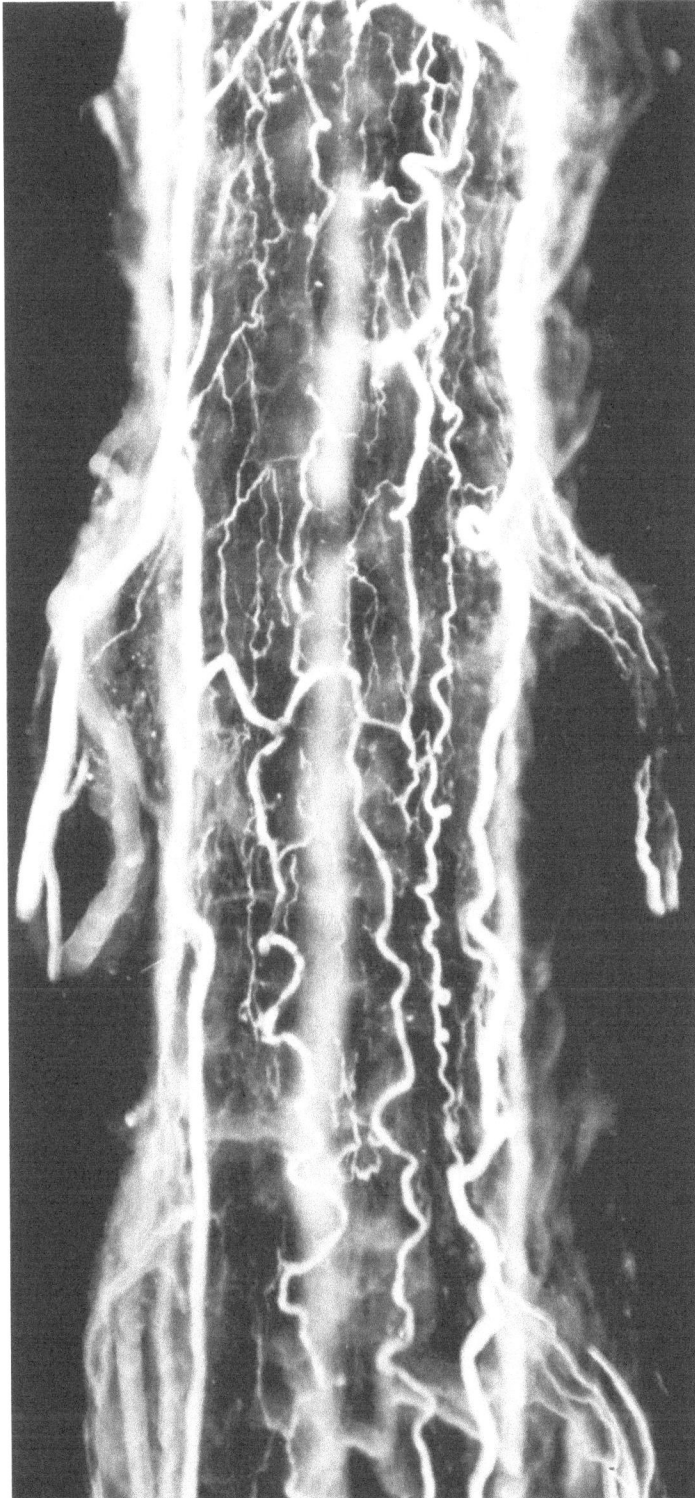

figure 2.7

A detailed photograph of the posterior
surface of the spinal cord illustrated
in Figure 2.4. The *posterolateral
longitudinal arterial trunks* are clearly
shown with the pial plexus between
them. In the midline, the *anterior
median longitudinal arterial trunk*
can be seen out of focus. There are
four nerve roots visible on this
segment of the specimen. One large
posterior radicular artery is visible
on the left in the upper half of the
photograph. Opposite this there is
a well-defined loop in the right
posterolateral longitudinal arterial
trunk. The spatial relationships of
the fine arteries on the nerve root
at this level are shown in detail in
Figure 2.7. In the lower half of the
photograph a fine radicular artery is
clearly visible joining the *right
posterolateral longitudinal arterial
trunk* of the cord. Its corresponding
companion on the left side is less
clearly seen.

figure 2.8

A detailed view of the *right
posterolateral longitudinal arterial
trunk of the spinal cord* illustrated
in Figure 2.7 (higher magnification).
One of the fine radicular arteries on
the adjacent nerve root is seen
joining the *posterolateral longitudinal
arterial trunk* of the spinal cord.

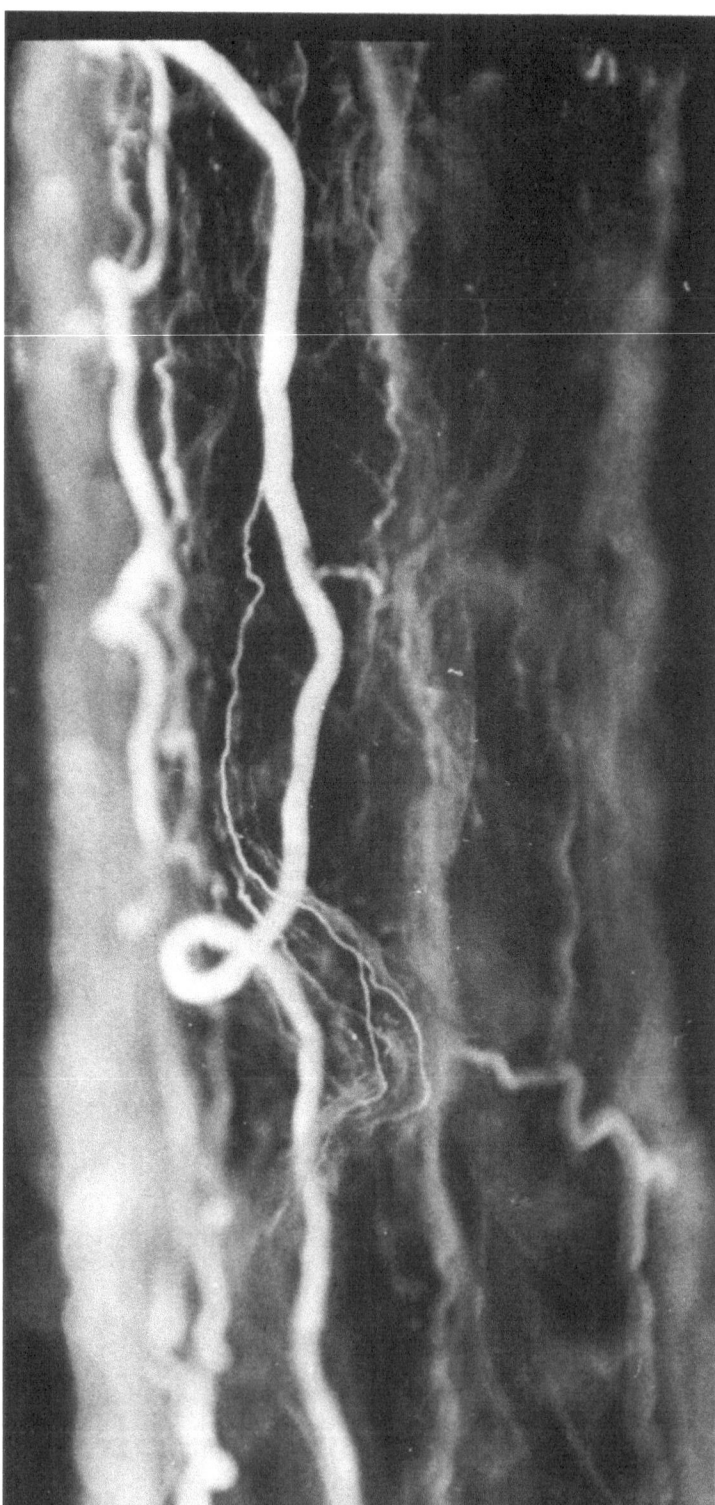

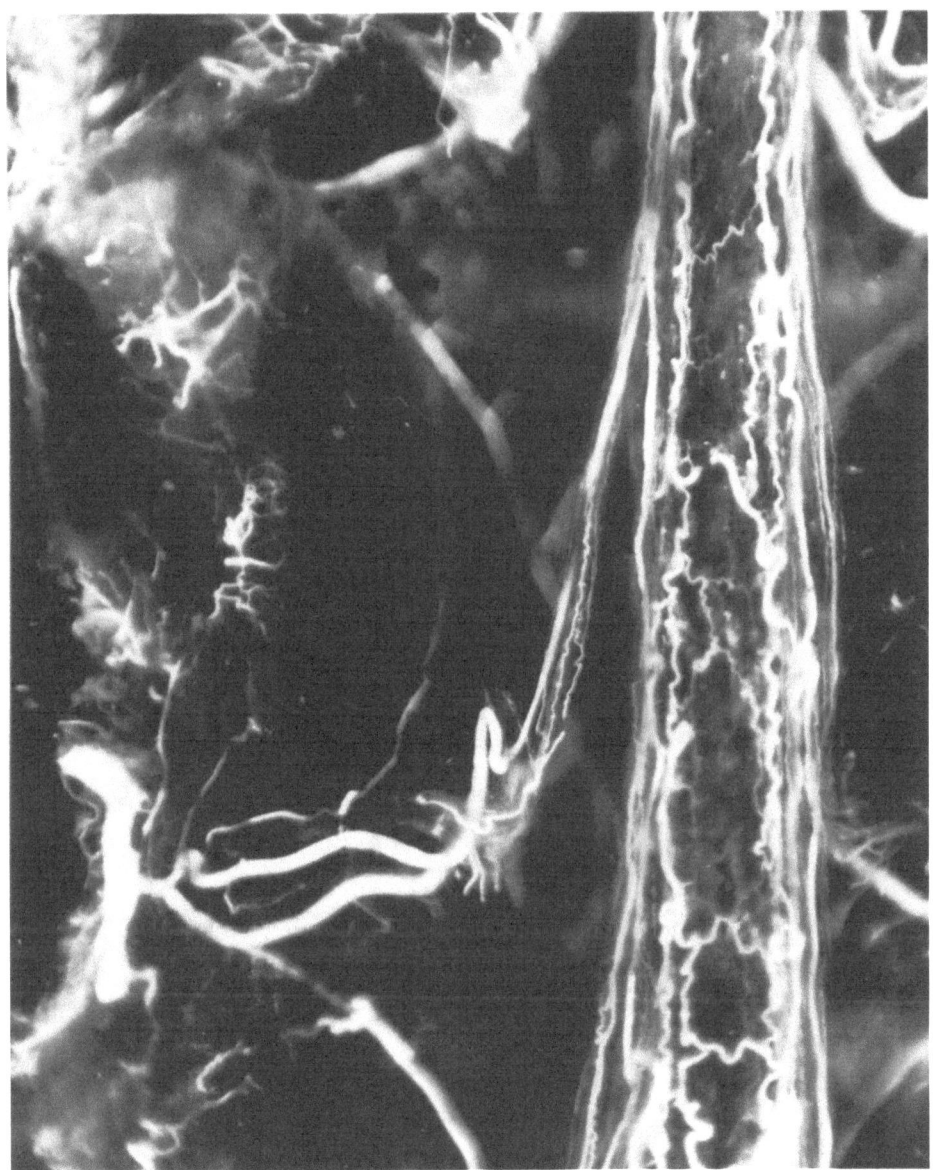

figure 2.9

A detailed photograph of the posterior surface of the thoracic spinal
cord of a female aged 35 years. This cord has been fixed by the
Spalteholz method and dissected *in situ*. The origin of a posterior
nerve root artery on the left side can be seen arising in common with
the anterior spinal branch of the intercostal artery. This vessel can be
traced from its origin, through its point of entry into the dural nerve
root sheath and then upward inside the dural sac until it reaches *the
posterolateral longitudinal arterial channel of the spinal cord*, with
which it anastomoses. It is accompanied by fine dural branches.

figure 2.10

A photograph of the anterior aspect
of the lower portion of the spinal
cord from a male child aged 12
years. The origin of a large anterior
radicular artery outside the spinal
canal can be seen. It courses
upward to the anterior surface of
the lower portion of the spinal cord,
where it bifurcates into ascending
and descending branches. It is
important to appreciate that this
specimen has been prepared
especially to demonstrate the course
of this individual vessel and to show
its relationship to *the anterior median
longitudinal arterial trunk of the
spinal cord.* Hence no other valid
inference can be drawn from the
examination of this specimen
relevant to the question of segmental
blood supply of the cord.

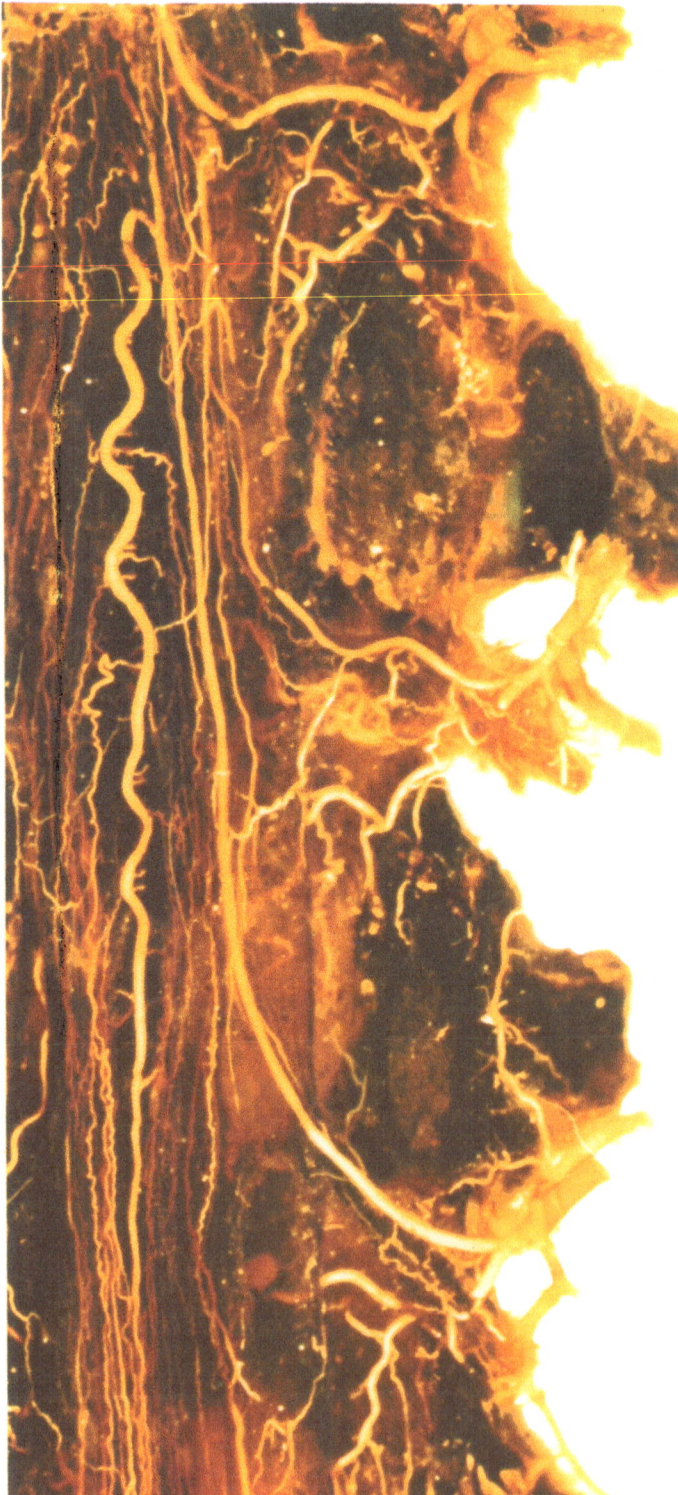

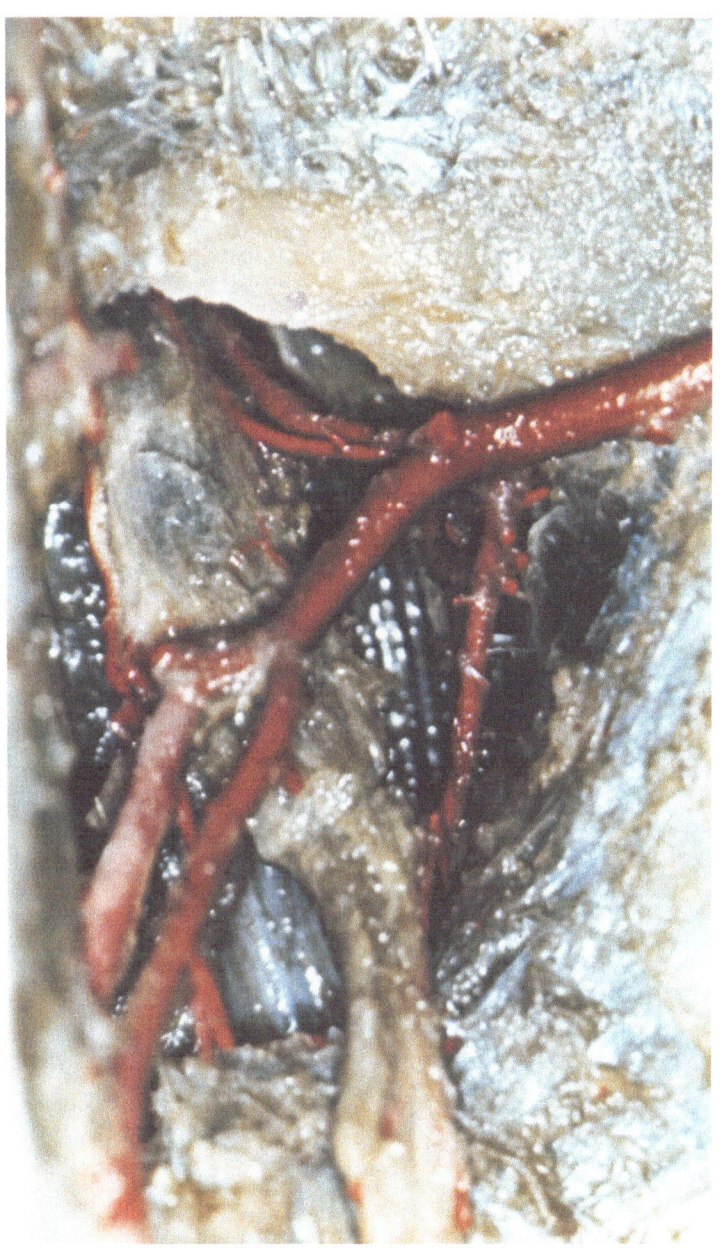

figure 2.11

A detailed photograph of the distribution of latex-filled arteries at the
level of L2 on the left side from the spine of a female aged 18 years.
The bifurcation of the anterior spinal canal branches of the lumbar
artery can be seen on the right of the specimen. A little further
posteriorly, arising from the upper border of the main stem of the
lumbar artery as it courses backward, the anterior and posterior
radicular arteries can be seen. They run upward along the nerve root
before penetrating the dural sleeve, just medial to the vertebral
pedicle.

figure 2.12

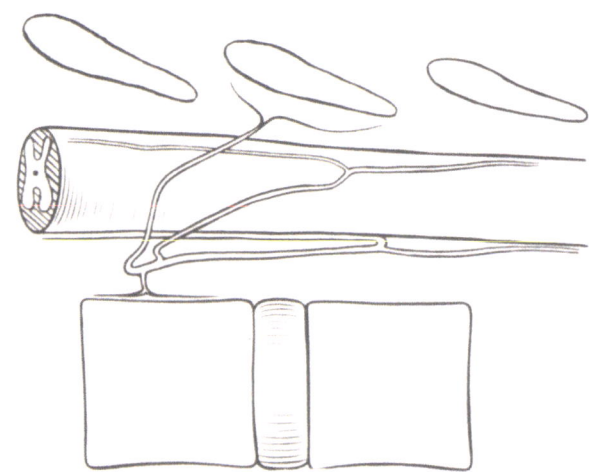

A photograph of the thoracic portion
of the spinal cord from a young
adult following arterial injection and
Spalteholz preparation of the
specimen. The cord is seen lying in
the spinal canal from which the
lateral walls have been dissected
free. The course of the neural
branches on two adjacent nerve
roots can be seen, as indicated in
the accompanying simplified
schema. In this preparation the dural
sac is intact, but transparent.

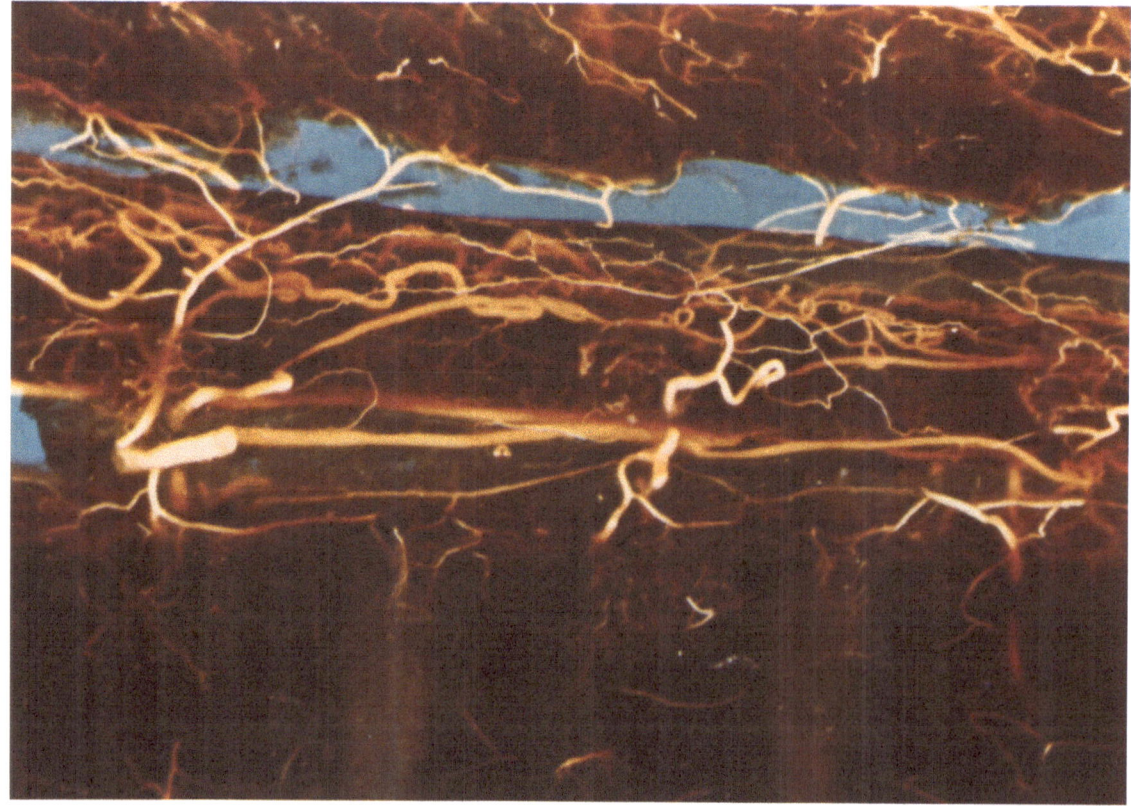

figure 2.13

A photograph of the posterior surface of the conus medullaris and
cauda equina from the spine of a female aged 18 years. *The right
and left posterolateral longitudinal arterial trunks of the spinal cord*
can be seen converging at the inferior point of the conus medullaris.

figure 2.14

A photograph of the tip of the conus
medullaris, showing the confluence
of *the right and left posterolateral
longitudinal arterial trunks of the
spinal cord* with the tortuous *anterior
median longitudinal arterial trunk*
in a male child of 17 months.
Spalteholz cleared specimen,
photographed by transmitted light,
magnified approximately by 5.

figure 2.15

A photograph of the tip of the conus
medullaris from a male child aged 17
months viewed from the lateral
surface. The junction of the *anterior
median longitudinal arterial trunk*
with the *posterolateral longitudinal
arterial trunk of the spinal cord* is
shown, the latter being on the right
of the photograph. Tortuous pial
anastomoses between both systems
are seen.

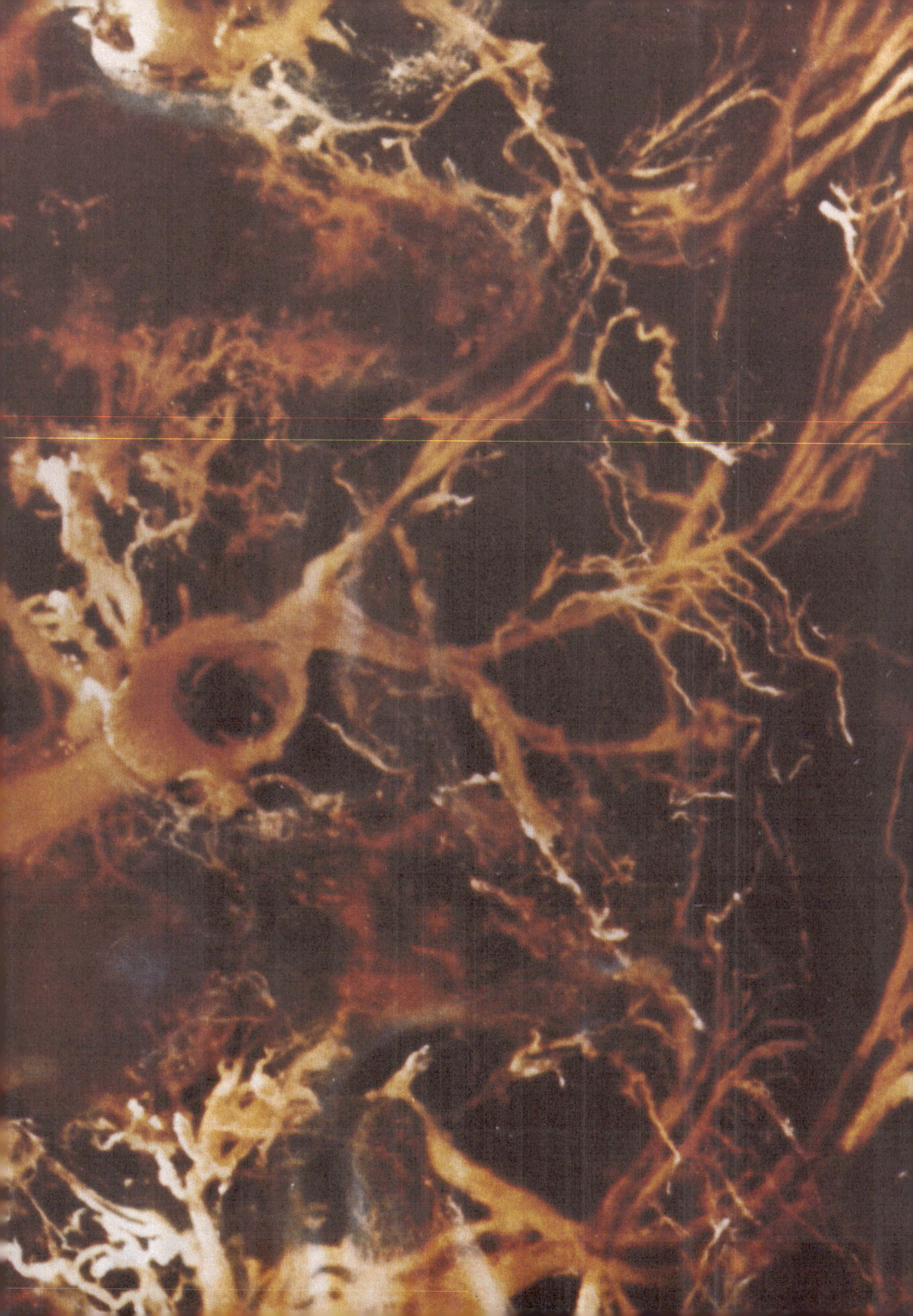

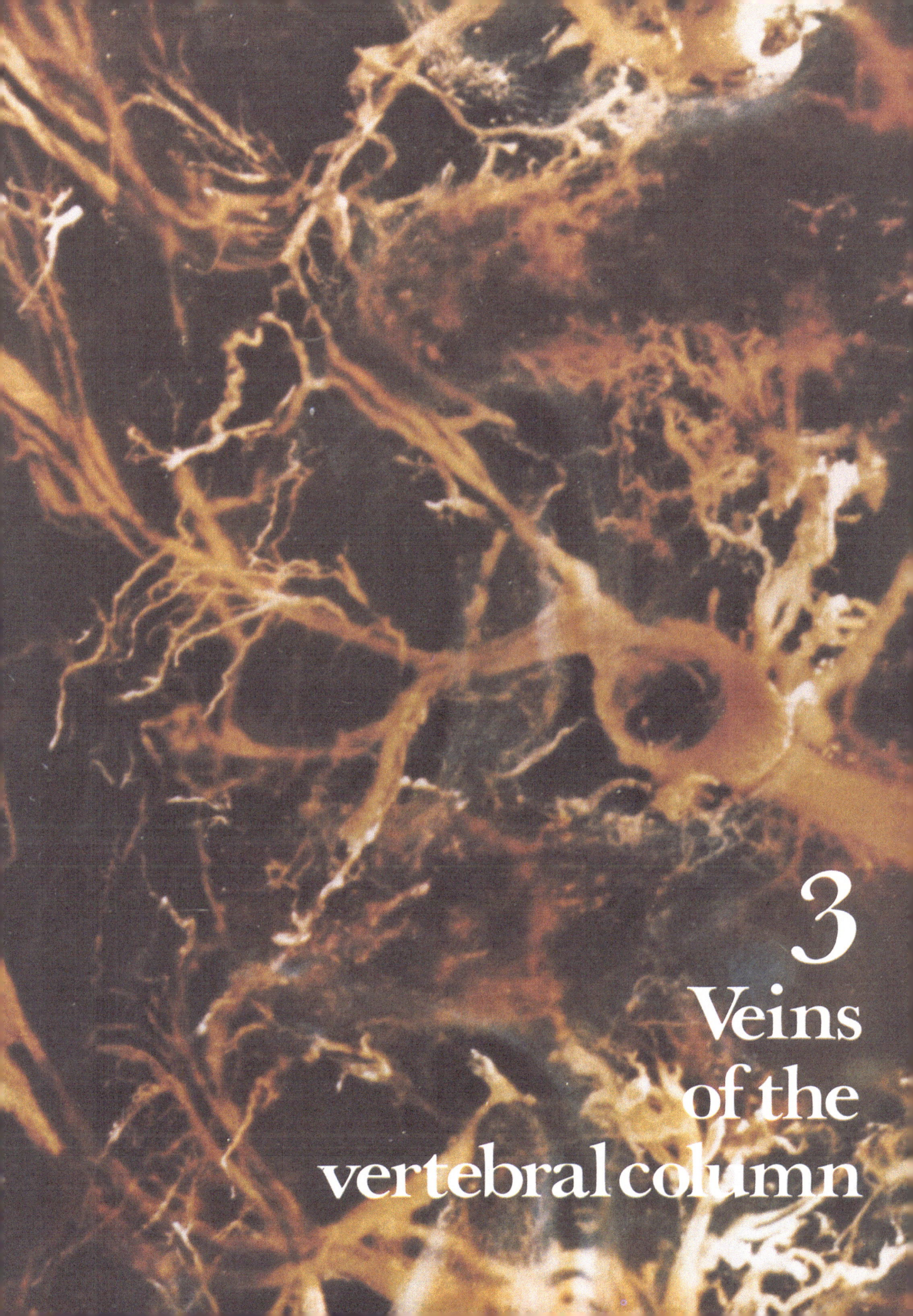

3
Veins
of the
vertebral column

In 1947 William Herlihy, an Australian anatomist, published a fascinating study entitled: "Revision of the Venous System: The Role of the Vertebral Veins." His theme was to examine the anatomy of the spinal venous system and to relate it functionally to the dynamics of the circulation.

While the description of this complex venous system merits only a brief description in the 35th Edition of *Gray's Anatomy* (1973), Herlihy placed it in a new light in terms of its physiologic importance in the circulatory system of man. For this reason we have devoted considerable time to the preparation of specimens illustrating the outstanding anatomic features of these veins.

The vertebral column has two plexiform systems of veins extending along its length. These are described as the *external vertebral venous plexuses* and the *internal vertebral venous plexuses.*

THE EXTERNAL VERTEBRAL VENOUS PLEXUS

The external vertebral venous plexus is subject to many variations in the spatial arrangements of its tributaries. Described in simplest terms, its main posterior branches form large veins which relate themselves to the sides of the spinous processes and course forward across the laminae on each side toward the intervertebral foramina (Fig. 3.1). They coalesce outside the foramina with emerging intervertebral veins. Here they are joined also by anteriorly directed veins which drain the body wall, the confluence forming segmental veins related to the sides of the vertebral bodies. These veins in the neck drain largely into the venae comitantes of the vertebral arteries in front, while the posterior tributaries of this plexus join the deep cervical veins on each side. In turn, both these veins and the right and left vertebral veins drain into the brachiocephalic veins near the internal jugular entry points. In the thoracic region they are termed intercostal veins which, further anteriorly, join the azygos systems of longitudinally directed veins of large caliber. In the lumbar region they are known as the lumbar veins, corresponding to the named arteries in this region, but are connected by a variable series of longitudinally directed channels, the *ascending lumbar veins* and the *lumbar azygos veins.*

The lower lumbar veins drain into the inferior vena cava while the azygos systems in the chest drain into the superior vena cava and left brachiocephalic veins. From a functional point of view, there are important anastomoses between the external vertebral venous system and certain of the visceral veins such as the pelvic plexus and the renal veins.

THE INTERNAL VERTEBRAL VENOUS PLEXUS

This complex system of veins extends from the region of the sphenoidal clivus within the skull, where it anastomoses with the sinuses at the base of the skull, to the sacral region below. It is in two parts, the *anterior internal vertebral venous plexus* and the *posterior internal vertebral venous plexus.* Batson (1940) described the importance of its continuity with the prostatic plexus and noted its large capacity. When intraabdominal pressure is high, venous blood from the pelvic plexus passes upward in the internal vertebral venous system. Likewise, when the jugular veins are obstructed, blood leaves the skull via this plexus.

Both the anterior and posterior internal vertebral venous plexuses are arranged in an arcuate pattern overlying the sharply defined arterial arcades described in Chapter 1. The plexuses surround emerging nerve roots at the level of the intervertebral foramina and just outside them fuse with the segmental veins of the external vertebral venous plexus.

The connections of the internal vertebral venous plexus with the intrinsic veins of the vertebral body are described in greater detail in Chapter 6.

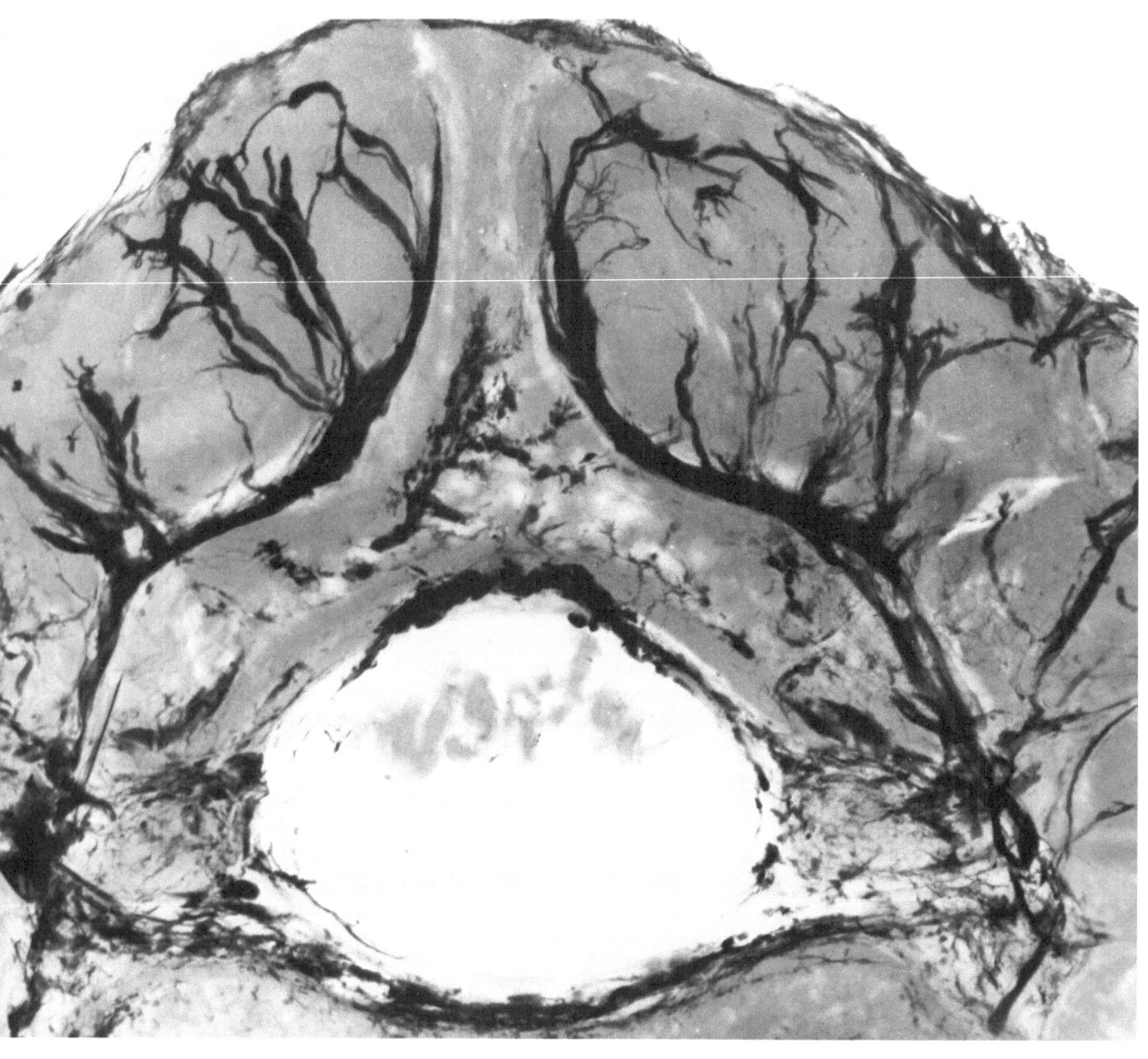

figure 3.1

A radiograph of a transverse section of a lumbar vertebra from an adult
to show the relations of the main stems of the lumbar veins (posterior
external vertebral venous plexus) to the spinous process and lamina.
(Reproduced by courtesy of J. B. Lippincott and Company from *Clinical
Orthopaedics and Related Research*, No. 115, 1976.)

figure 3.2

A photograph of a dissection to show
the external vertebral venous plexus
in the thorax from a male aged 26
years.

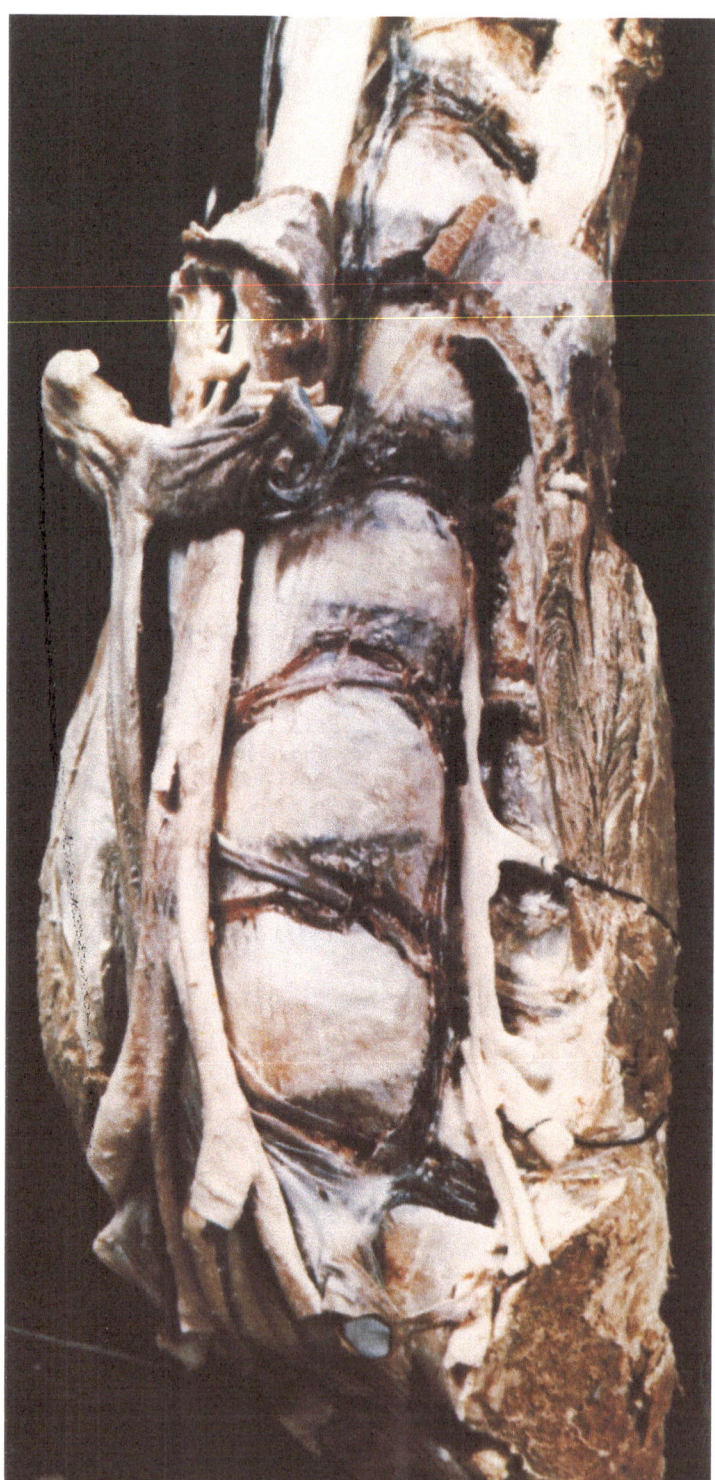

figure 3.3

A photograph of the specimen
illustrated in Figure 3.2, showing
the veins related to the lower half
of the vertebral column viewed
obliquely from the left side.

figure 3.4

A photograph of the specimen
illustrated in Figure 3.2, viewed
from the left side.

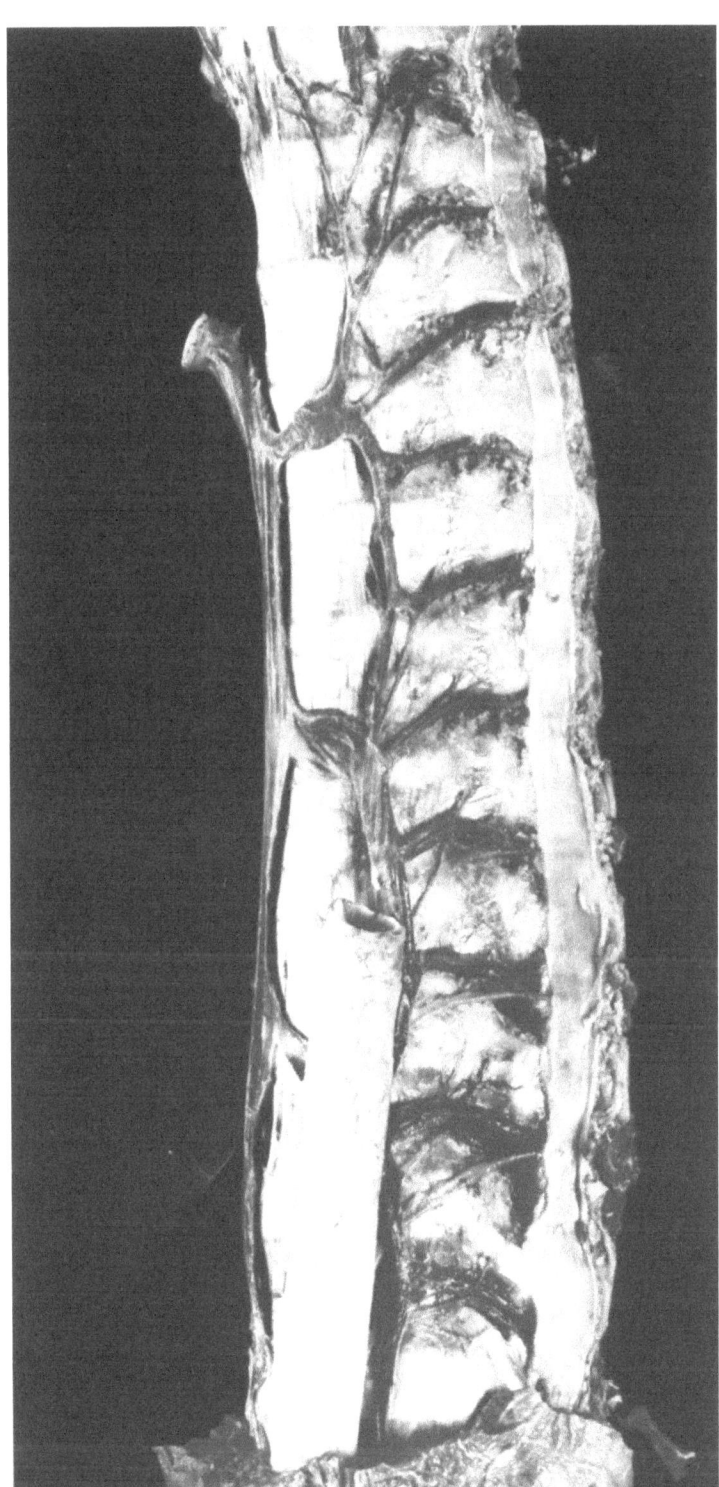

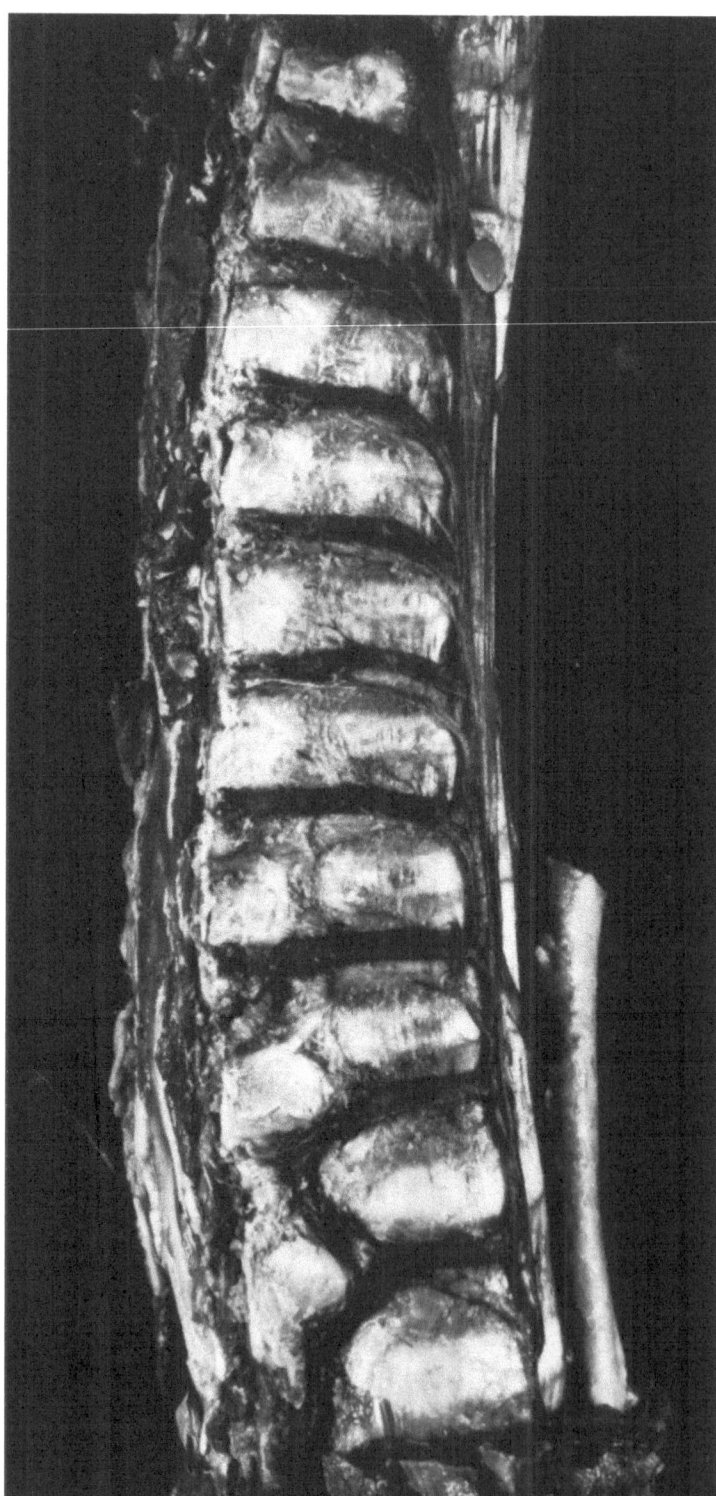

figure 3.5

A photograph of the specimen
illustrated in Figure 3.2, viewed
from the right side.

figure 3.6

A photograph of the anterior surface
of the spinal canal in the upper
lumbar region from an adult showing
the internal vertebral venous plexus
filled with blood. The dura mater has
been left intact and rendered
transparent. Note the arcuate
pattern of the plexus with the
right- and left-sided arcs meeting
in the middle of each vertebral
body.

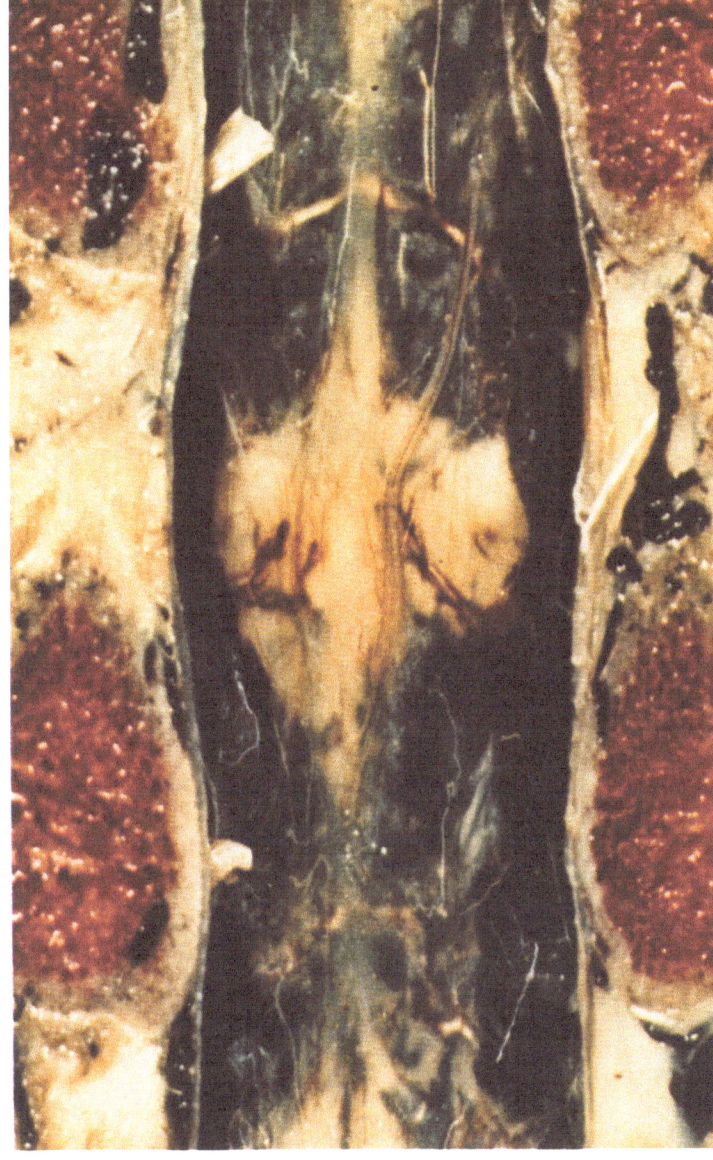

figure 3.7

A photograph of a dissection of the
anterior internal vertebral venous
plexus in a male aged 66 years.
The posterior longitudinal ligament
has been removed. The plexus was
injected with epoxy resin.

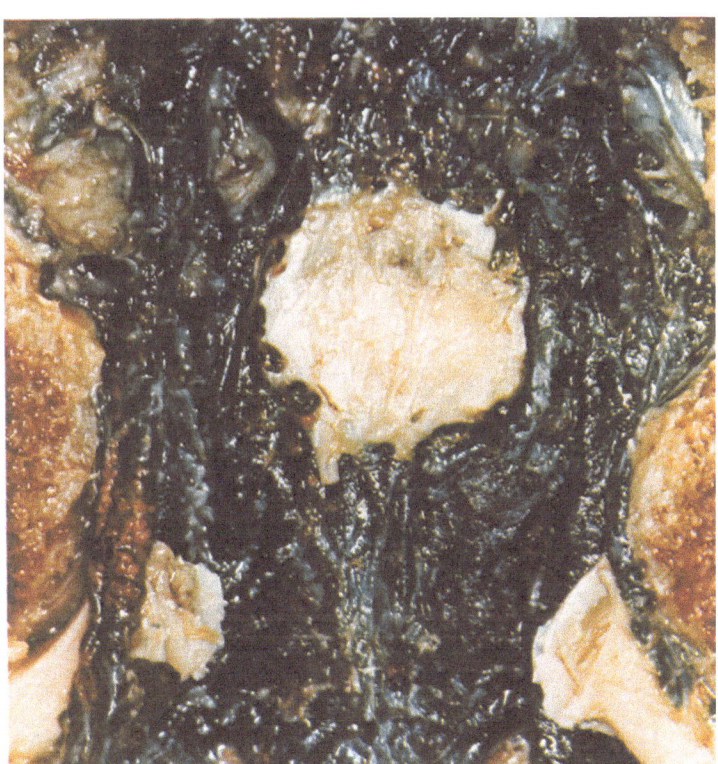

figure 3.8

A photograph of a dissection to show details of the anterior internal
vertebral venous plexus from a male aged 71 years. The plexus was
injected with latex rubber. Note the large radicular vein joining the
plexus.

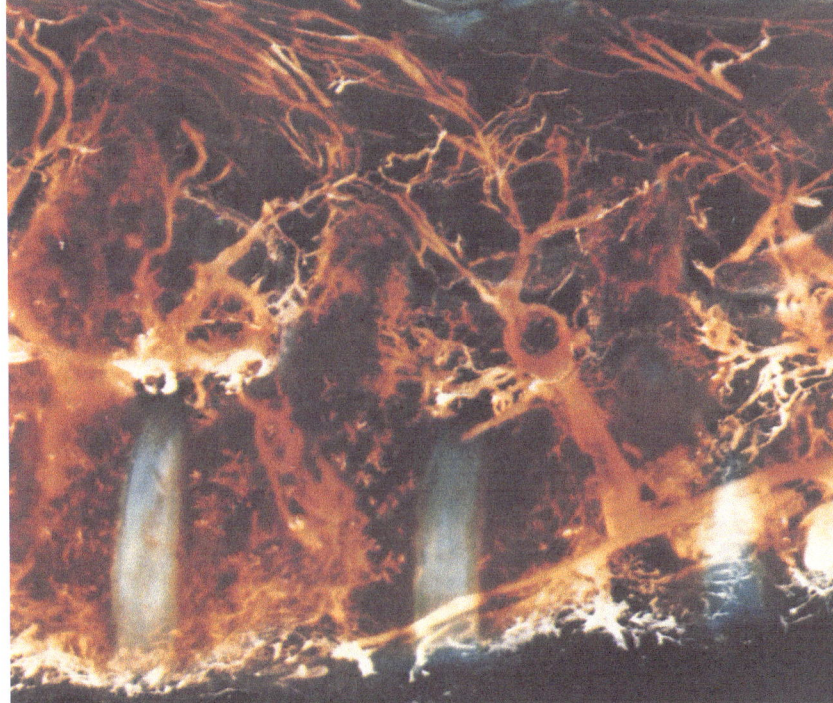

figure 3.9

A photograph of a thin sagittal
section of the spine of an adult cut
laterally at the level of the pedicles
and viewed from the medial
(spinal canal) side. This is in fact
a view of the intervertebral foramina
looking from within to the outside.
The vertebral veins have been
injected with Micropaque. Note the
pedicle veins which enter the inferior
surfaces of the intervertebral veins.

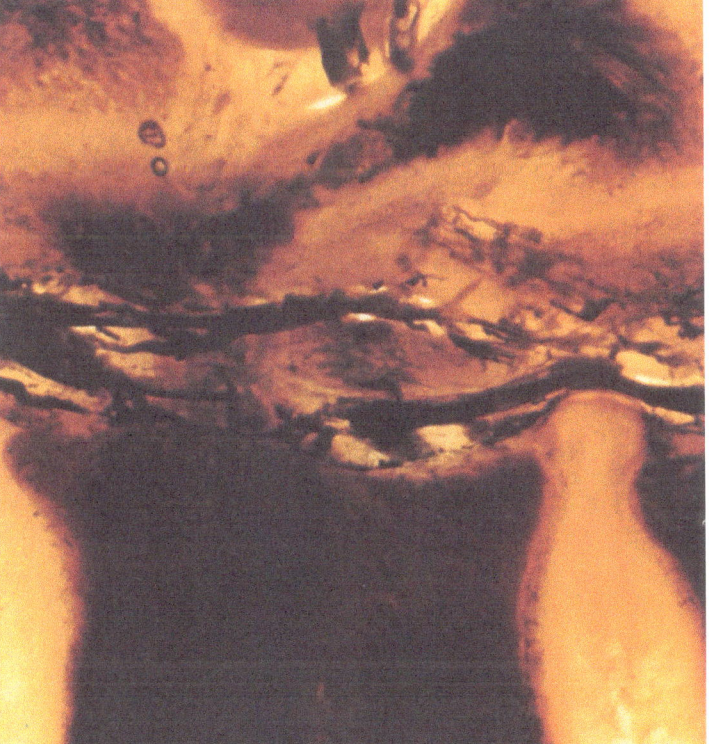

figure 3.10

A detailed photograph of a thin
sagittal section of a lumbar vertebra
from an adult (Spalteholz cleared
specimen, photographed with
transmitted light). The anterior
and posterior internal vertebral
venous plexuses are shown inside
the spinal canal at the level of the
vertebral pedicle and surrounding
the emerging nerve root in the center
of the specimen.

4
Veins
of the
spinal cord

Of all the veins in the body, those on the surface of the spinal cord have been most difficult to study. In life they are transparently thin and at postmortem when collapsed, they are difficult to identify and fragile to handle. While most authors would agree that satisfactory filling of the arterial blood vessels of the spinal cord can be achieved only by injecting the aorta in the intact body (Gillilan, 1958), little success has been reported with venous injections. In this chapter we present information on the surface veins of the spinal cord obtained by two methods of injection in the intact body: In the first, following arterial injection we have processed some specimens by Spalteholz's clearing method excluding bleaching agents. In this way arteries have been shown filled with Micropaque (white) and the veins are filled with blood (black) (Figs. 4.1, 4.3). In the second, retrograde filling of veins by injection through the azygos system as described on page 119 has led to complete filling of the large surface veins.

The surface veins of the spinal cord conform to the basic outlines of the arteries, with the exception that there is a well-defined dorsal median longitudinal venous trunk which has no named arterial counterpart. In the adult the veins are large and tortuous, their wavy courses tending to confuse their underlying simple architectural plan.

Illustrations have been selected to demonstrate the salient features of this venous anatomy, with detailed descriptions in the accompanying legends.

figure 4.1

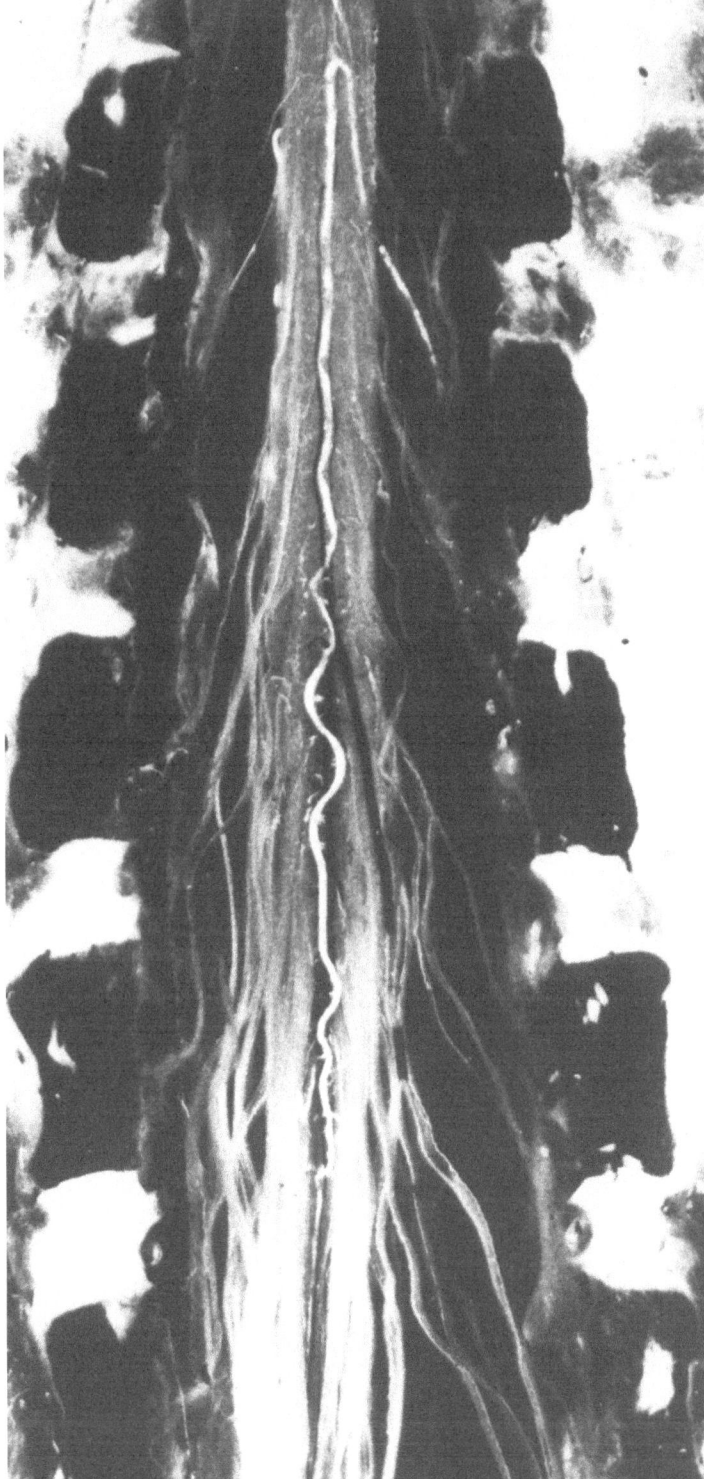

A detailed photograph of the anterior
surface of the distal third of the
spinal cord in a child aged 6 years.
The arterial system had been injected
with Micropaque and some of the
spinal cord veins are shown filled
with blood. The specimen has been
prepared by fixation and Spalteholz
clearing of the intact vertebral
column, followed by careful
dissection with removal of the
vertebral bodies from the front of
the spinal canal. The cord has been
illuminated with incident light. The
anterior median longitudinal artery
of the spinal cord is seen overlying
the centrally placed anterior
longitudinal venous trunk of the
cord. Note the large medullary vein
on the right side of the photograph
which is partly filled with clotted
blood.

At the top of the specimen on the
right side, a large tributary of the
anterior median longitudinal arterial
trunk of the spinal cord can be seen
(artery of Adamkiewicz). Note the
corresponding segmental artery on
the left side, which is minute by
comparison, but nonetheless present.

Note also that the anterior median
longitudinal venous trunk is
duplicated in the upper half of its
course in this specimen.

figure 4.2

A detailed photograph of the conus
medullaris viewed from in front with
the cauda equina spread out. Venous
injection. Male aged 21 years. The
cord was fixed *in situ*. Note the
large single anterior longitudinal
venous trunk over the lower portion
of the conus medullaris giving rise
to many tributaries which run along
the nerve roots of the cauda equina.
One particularly large channel on
the left side of the photograph
extends downward along the nerve
roots. Proximally in the specimen
the anterior longitudinal venous
trunk is duplicated.

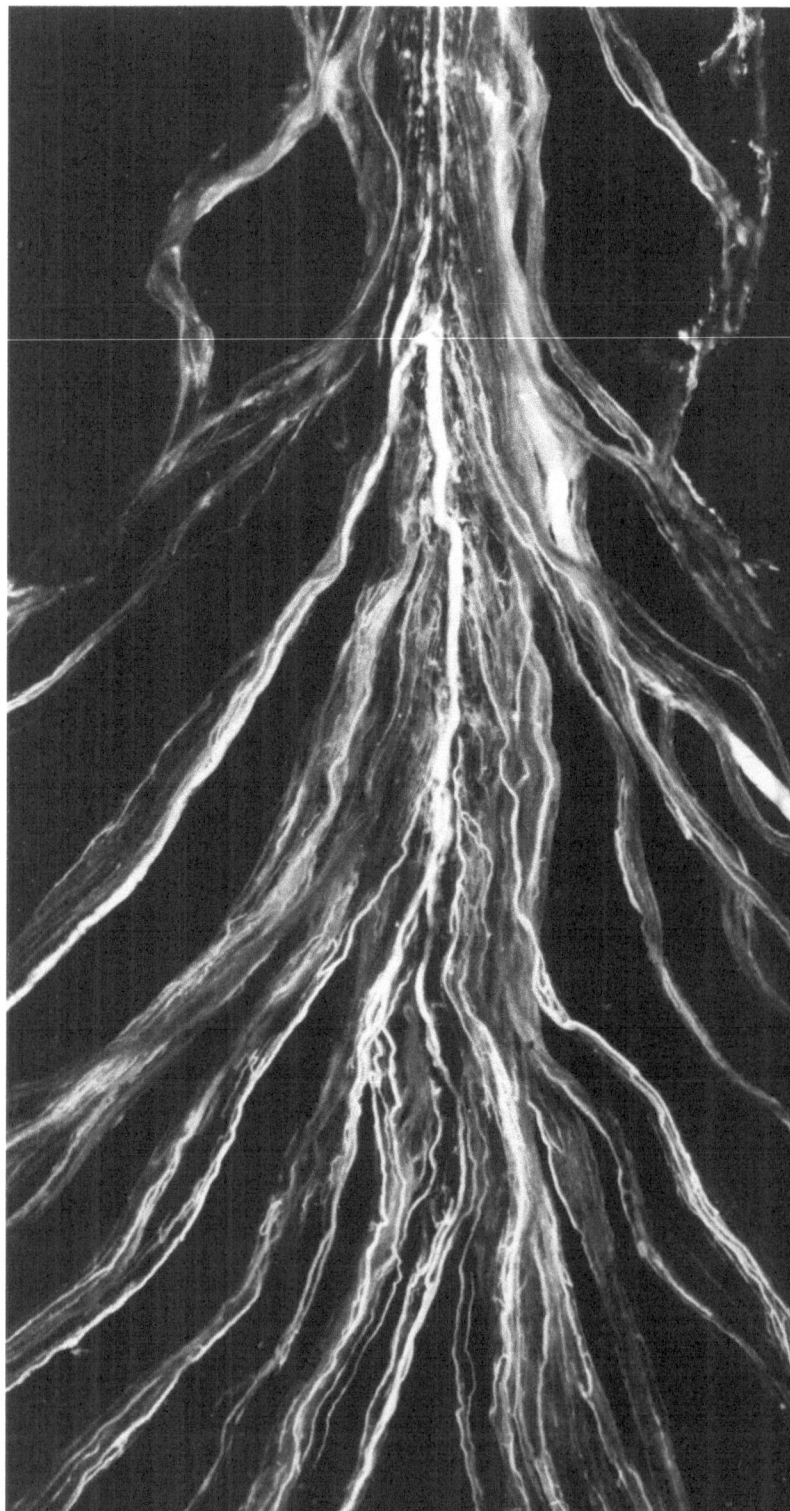

The blood supply of the vertebral column
and spinal cord

figure 4.3

A photograph of the posterior surface
of the lower half of the spinal cord
from a male child aged 6 years. The
arteries have been injected with
Micropaque in the intact body and
the veins are filled with blood. The
specimen was prepared by fixation
of the whole vertebral column and
processing by Spalteholz method
before dissection. The spinal canal
has been opened from in front and
behind and the bony remnants
separated to allow photography,
using transmitted light. There is a
posterior median longitudinal
venous trunk and in addition, parallel
venous channels corresponding to
the posterolateral longitudinal
arterial trunks of the spinal cord.
Note that the venous drainage is
strictly segmental though the size of
individual segmental veins varies.

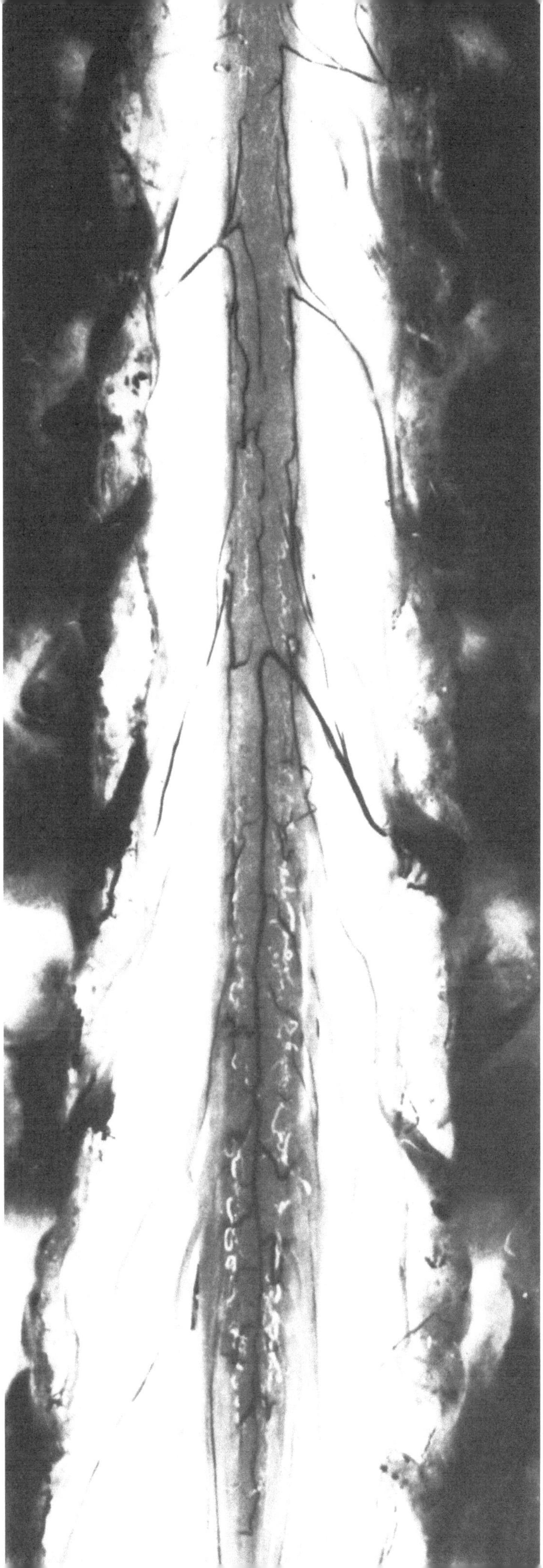

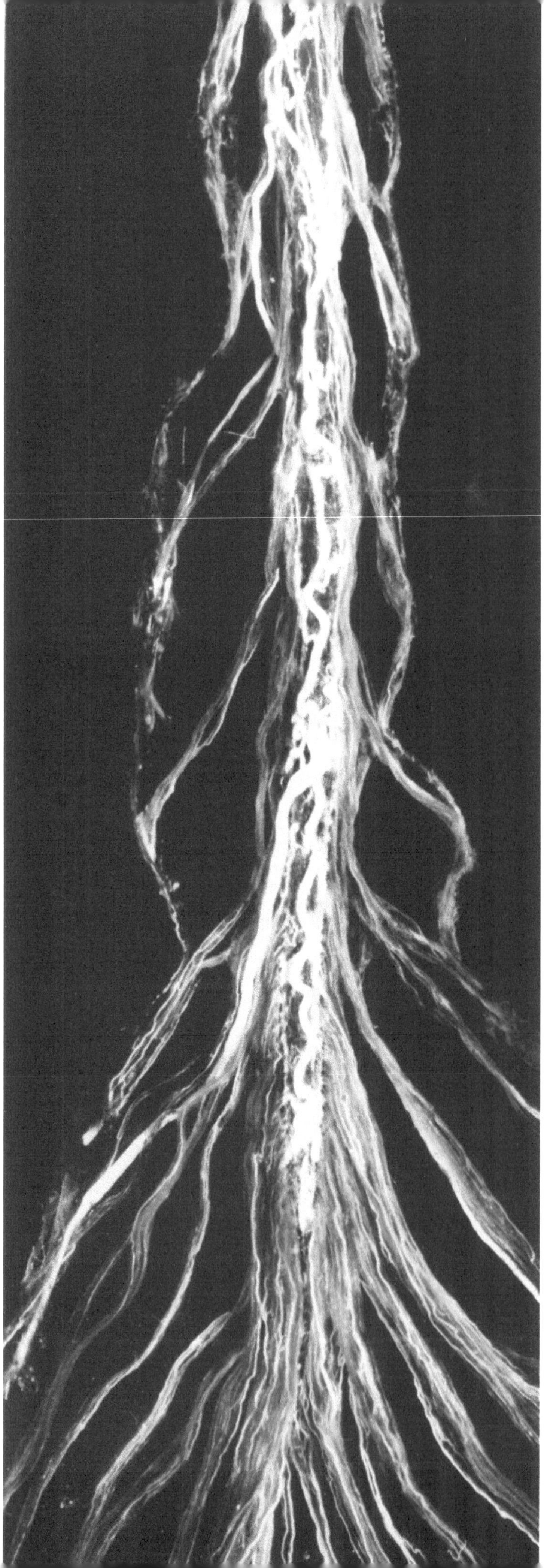

figure 4.4

A photograph of the posterior
surface of the spinal cord from a
male aged 21 years, prepared in the
same manner as the previous
specimen, but photographed with
incident light. In the adult, the
venous channels on the posterior
surface of the spinal cord are very
complex and more tortuous than are
the corresponding channels in the
child. However, the basic spatial
distribution of these channels is the
same as those shown in the child
(see Fig. 4.3). A number of large
branches leave the median vein.

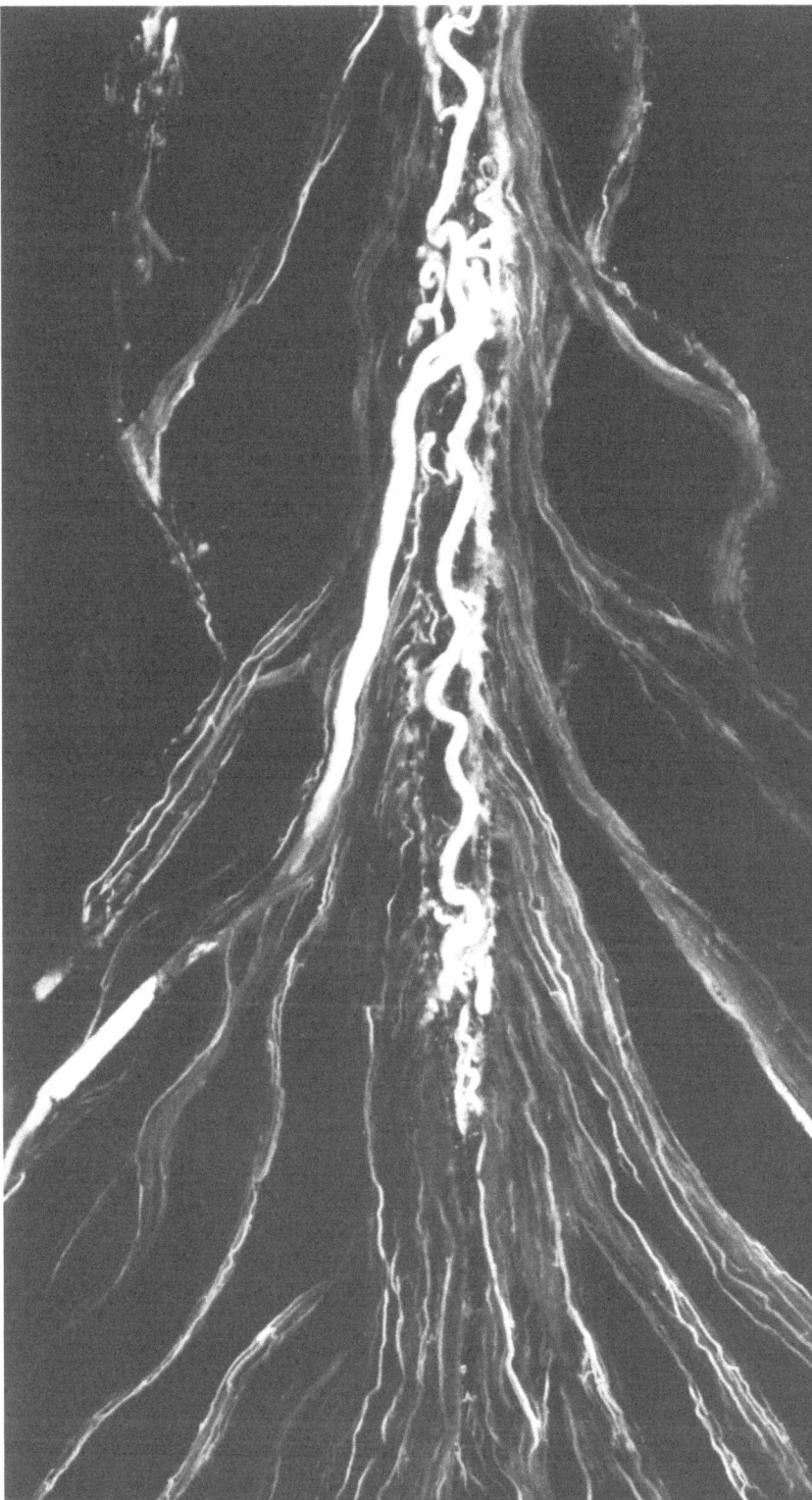

figure 4.5

A detailed photograph from the specimen illustrated in Figure 4.4, showing the arrangements of the posterior median longitudinal venous trunk over the conus medullaris. Note its tortuous course and the size of the vein which leaves it at a bifurcation on the left side of the photograph. Note also the veins which run parallel to the central channel, corresponding in position to the posterolateral longitudinal arterial trunks of the spinal cord. Each nerve root in the cauda equina is accompanied by a vein, which connects either with the median longitudinal venous trunk or the posterolateral longitudinal trunks.

figure 4.6

A photograph of the posterior
surface of the conus medullaris and
cauda equina from a male aged 71
years in whom the venous system of
the spinal cord had been injected
with latex rubber. In the upper part
of the photograph, the posterior
median longitudinal venous channel
of the spinal cord is seen and from
this, on the left side, a large, long
vein courses downward in company
with radicles of the cauda equina,
to penetrate the dural sleeve of the
second lumbar nerve root
inferomedially, where it joins with
tributaries of the internal vertebral
venous plexus.

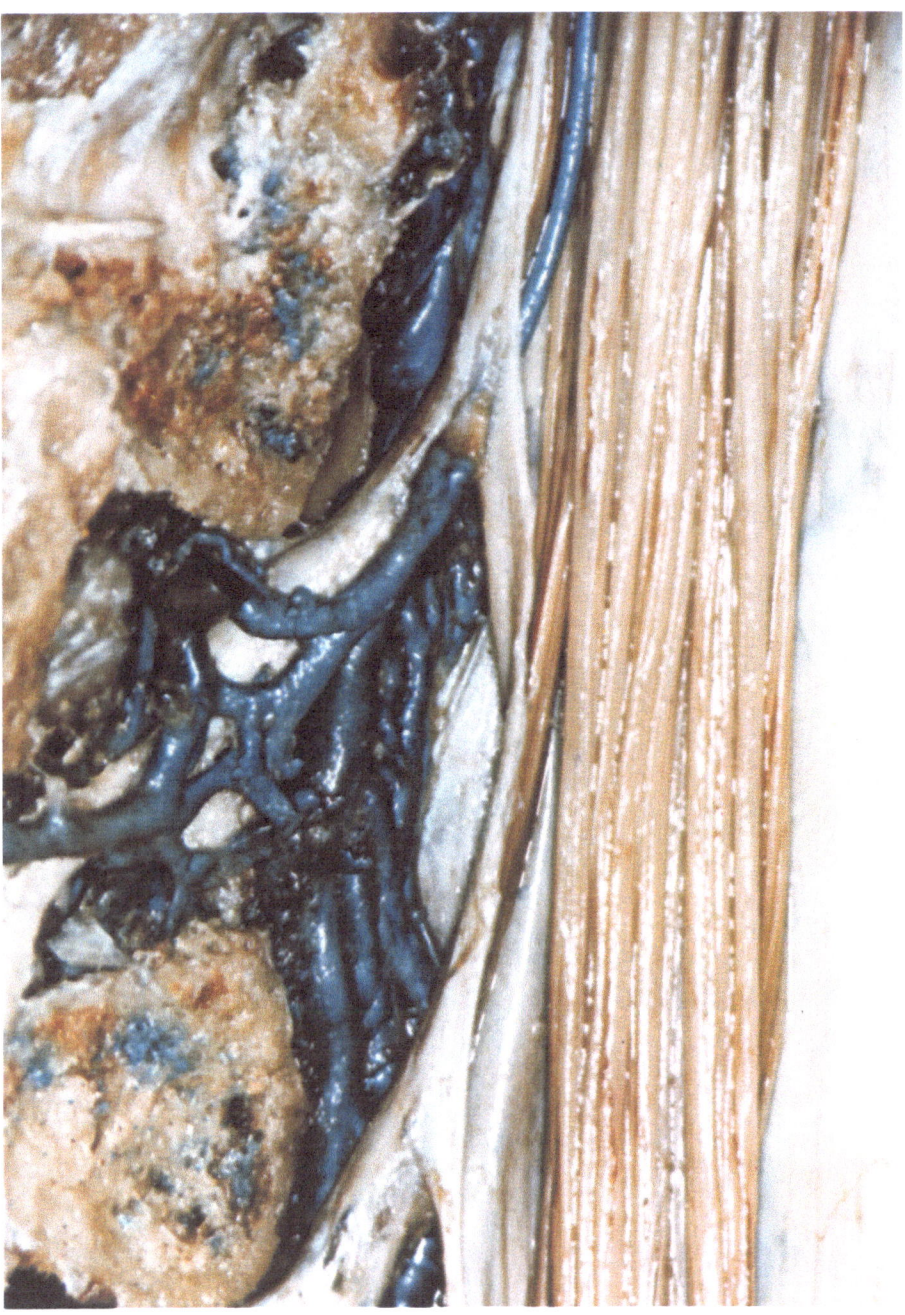

figure 4.7

A detailed photograph of the specimen illustrated in Figure 4.6, to
show the termination of the large radicular vein in the internal
vertebral venous plexus. Note the exit point of this radicular vein in
the axilla of the nerve root sleeve, some authors believe that the dura
exerts a valvelike effect on these veins (Dommisse, 1975).

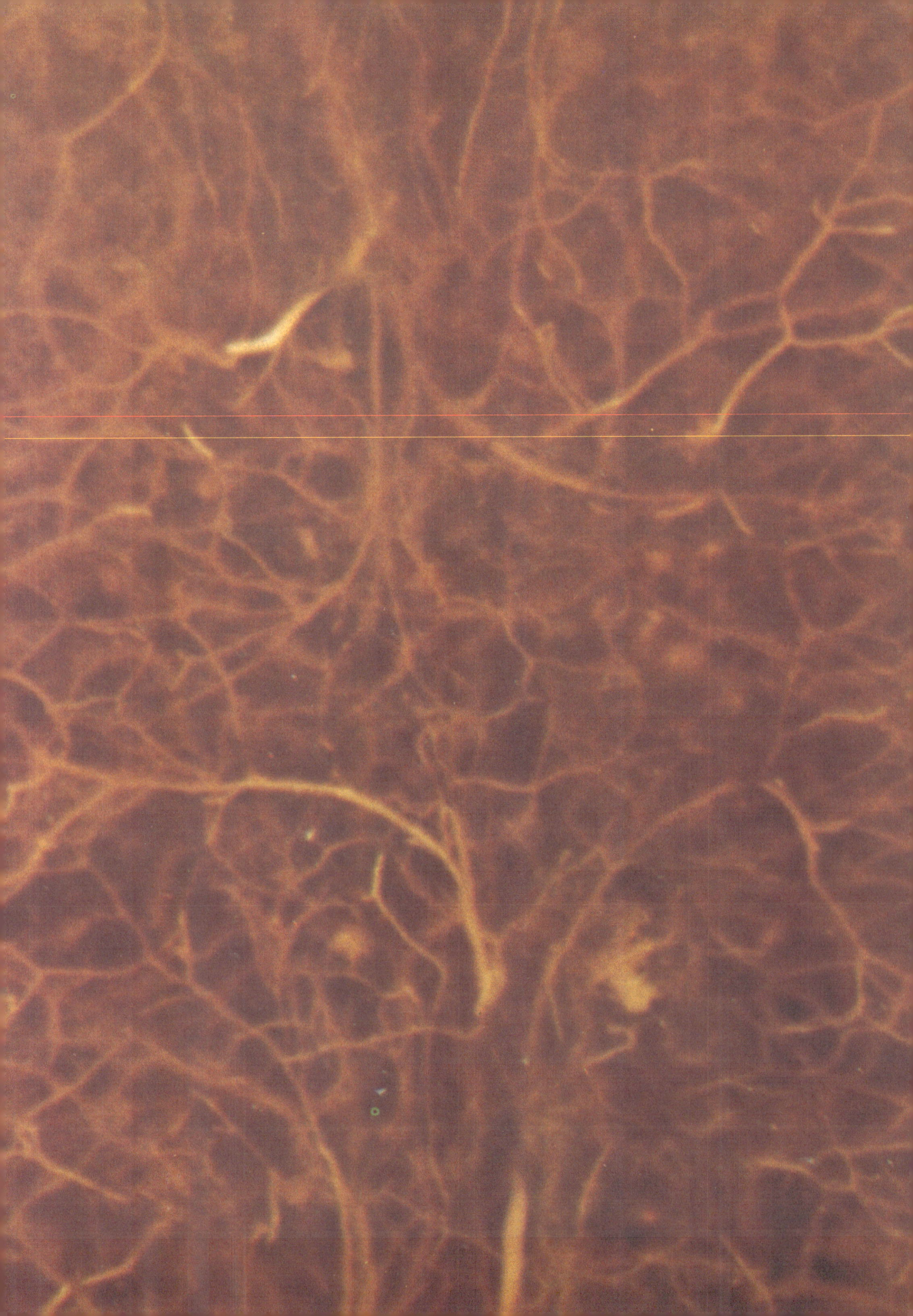

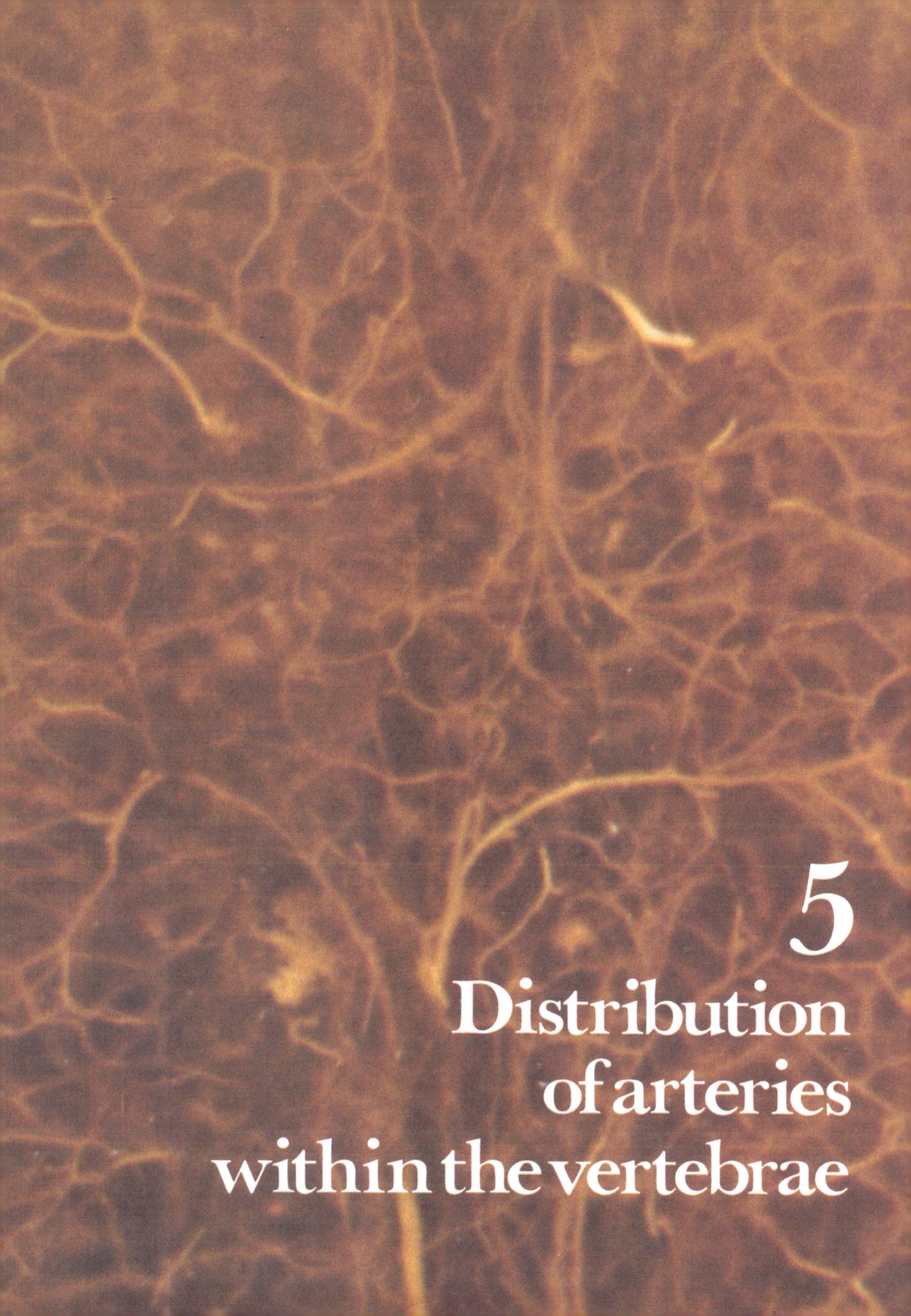

5
Distribution
of arteries
within the vertebrae

Within the interstices of the cancellous bone of the vertebral bodies lie dense and complicated arterial networks (Fig. 5.1). To study these vessels in detail the vertebral bodies should be divided into thin slices (0.5 to 0.75 cm) in three planes. It will be found that entrant vessels are arranged in constant basic patterns within the vertebral bodies, with most of their branches destined to be focussed on the vertebral end-plate zones.

In this chapter the distribution of arteries within the vertebrae from the various segments of the spine are illustrated. To avoid repetition, the description given in the text is of the intraosseous vessel distribution in lumbar vertebrae. Readers interested in other vertebral segments should therefore refer to the appropriate illustrations and their legends.

The centrum is penetrated radially in the horizontal plane by small arteries derived from the abdominal portions of the lumbar arteries anterolaterally and posteriorly by somewhat larger arteries derived from the arcuate branches of the anterior spinal canal division of the lumbar arteries (Fig. 5.2). An arterial grid is formed in the center of the vertebral body, from which vertical branches ascend and descend in slightly tortuous paths toward the respective vertebral end-plates, forming a brush border of arterioles which pass vertically into the vertebral end-plate cartilage capillary beds.

In coronal and sagittal sections (Figs. 5.3–5.8), the contributions from the ascending and descending branches of the lumbar arteries and analogous branches from the anterior spinal canal arcuate arteries to the vertebral bodies can be seen; these, too, have their entry points oriented circumferentially around the vertebral bodies. However, as the branches which enter the anterolateral aspects of the body penetrate to the interior, their main stems form triangular wedge-shaped patterns viewed in both the coronal and sagittal planes with apices near the junctions of the lateral and middle thirds or anterior and middle thirds of the vertebral body. From the sloping sides of these triangles, vertical branches turn upward or downward toward the vertebral end-plate areas. From the study of pure arterial injections it appears that the arterial grid in the centrum is concerned ultimately with the blood supply of the central third of the vertebral body and its respective vertebral end-plates. The remaining segments of the vertebral end-plates are supplied anterolaterally by the vertical branches arising from the triangular watershed described above and posteriorly by the ascending and descending branches of the arcuate arteries on the anterior wall of the spinal canal (Figs. 5.6–5.8).

The basic intraosseous patterns just described are established at birth and, as in other bones, remain unchanged throughout life except for certain details in the regions of the growing ends of the vertebrae. In the newborn child, the primary center of ossification for the body is surrounded superiorly and inferiorly by cartilage caps which merge with the intervertebral discs related to the upper and lower surfaces of the particular vertebral body. Already at this stage of development there is a central arterial grid which gives rise to a multiplicity of fine arterial branches spreading out in the directions described above. Some of these fine arteries penetrate the cartilage caps in the form of sinusoidal expansions (Fig. 5.9).

On the other hand, the cartilage caps are penetrated circumferentially by fine arteries branching from the ascending and descending branches of the lumbar arteries themselves. The terminations of these vessels in cartilage are clearly shown in a number of the accompanying illustrations (Figs. 5.9–5.12). These fine arteries narrow to arteriolar size and then branch out into sinusoidal terminations which are often Y-shaped with clublike endings. Small veins are formed within these sinusoids and they run back in the cartilage canals alongside the entrant arteries.

As the ossification of the vertebral body extends near to the final vertebral end-plate zone, so these sinusoidal systems disappear and in the adult a complicated vertebral end-plate capillary bed is formed with subarticular collecting vein systems which are described and illustrated in Chapter 6.

For practical purposes, the blood supply of the posterior vertebral elements has already been described in Chapter 1 (Fig. 1.12). The essential details of the arterial supply of sacral cervical and thoracic vertebrae are illustrated in Figures 5.15–5.23.

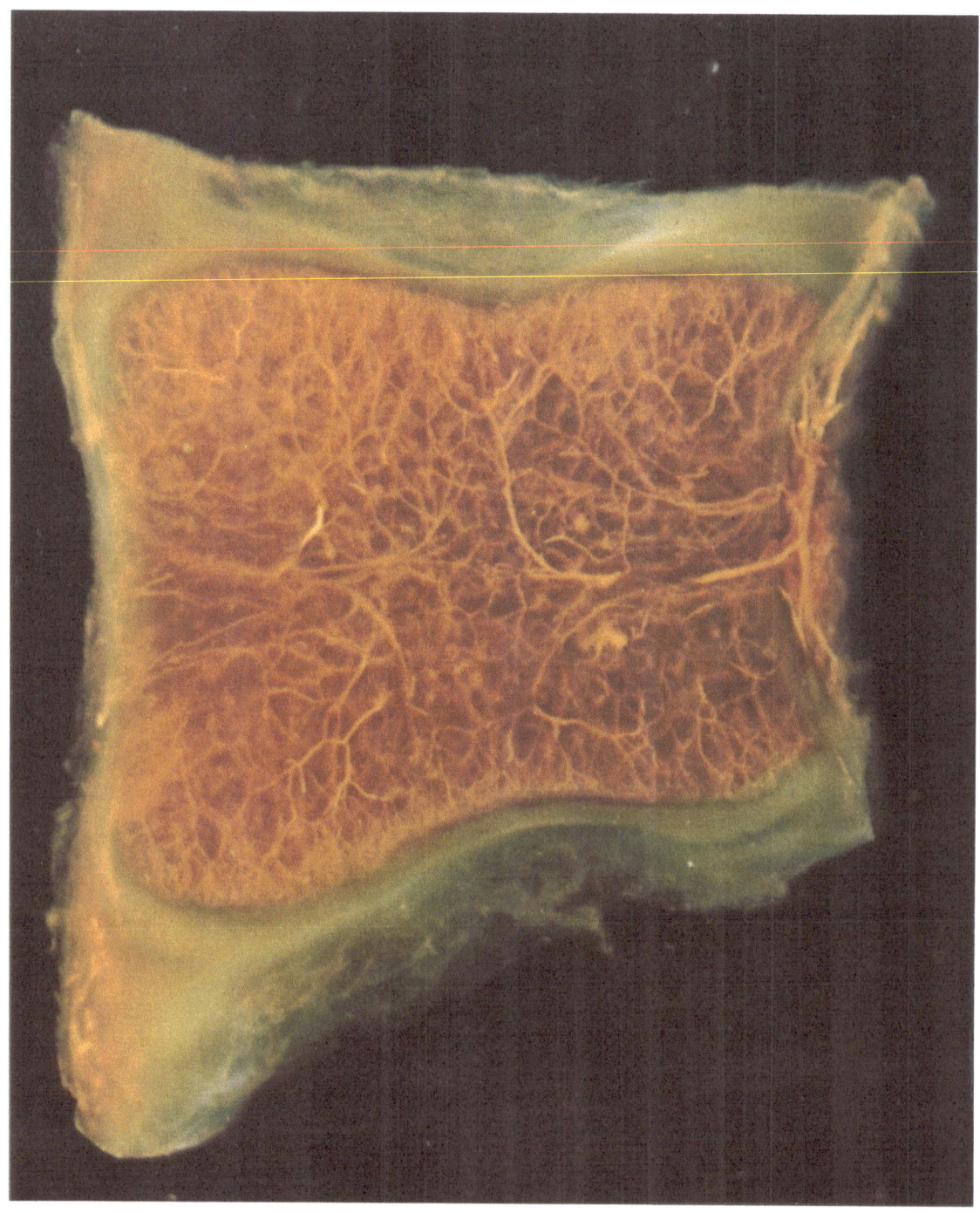

figure 5.1

A midsagittal section from the fifth lumbar vertebral body of a child
aged 15 years, to show the density of intraosseous arterial branching
within a normal vertebral body.

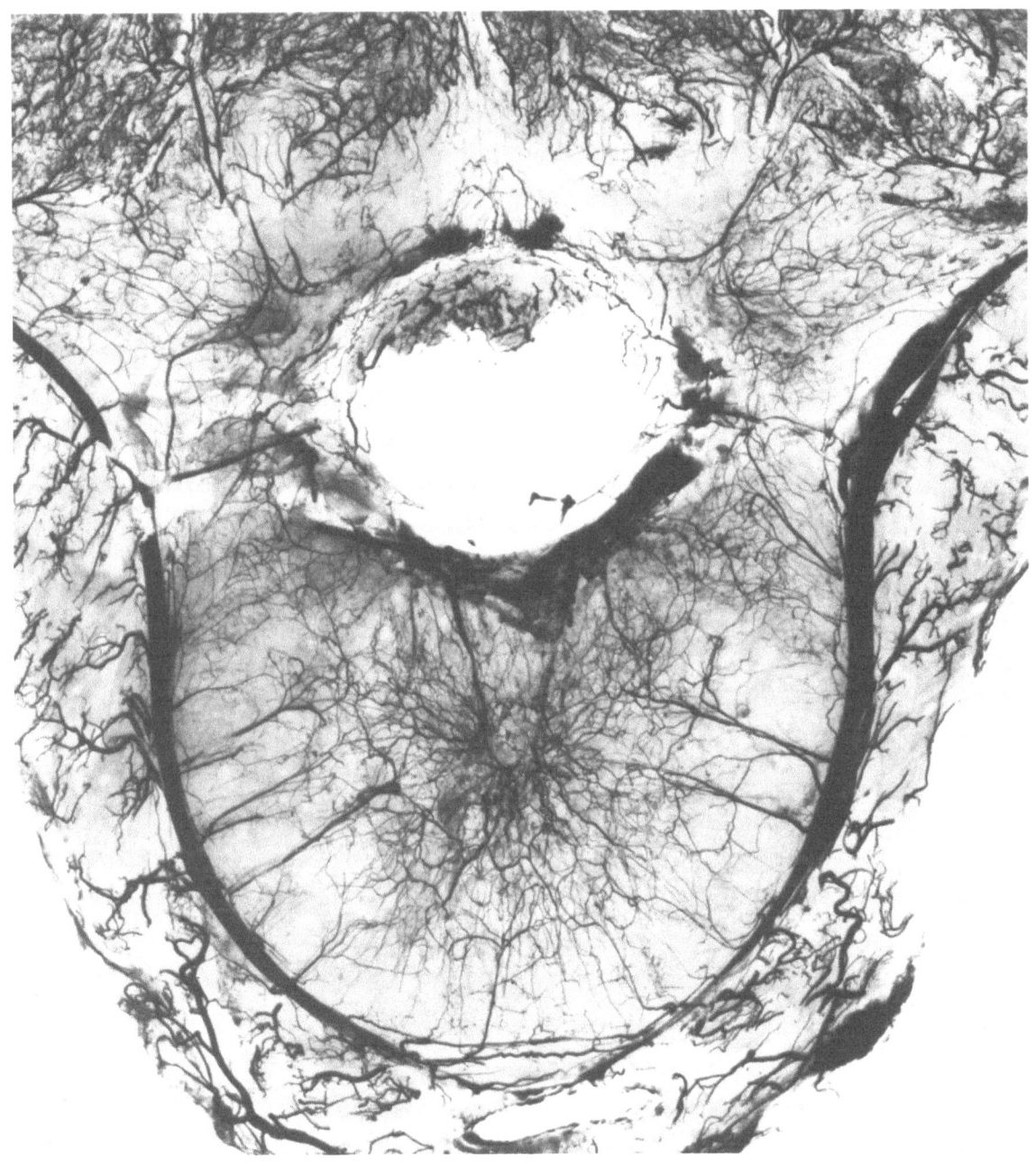

figure 5.2

A transverse section through the lumbar vertebral body of a child aged
13 years. The radiate distribution of centrum branches arising from the
inner surface of the lumbar arteries on each side has been shown. Note
the muscular branches passing directly into the muscles from the outer
side of the main trunk of each lumbar artery. (Reproduced by courtesy
of J. B. Lippincott and Company from *Clinical Orthopaedics and
Related Research*, No. 115, 1976.)

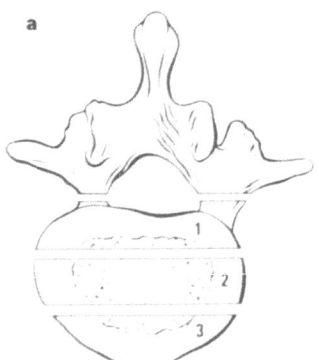

figure 5.3a and b

(a)
Line drawing to indicate the coronal planes of section of a lumbar vertebral body. (Reproduced by courtesy of J. B. Lippincott and Company from *Clinical Orthopaedics and Related Research*, No. 115, 1976.)

(b)
A coronal section of the second lumbar vertebra from a male aged 36 years corresponding to section 2 in the line drawing (b). On the left of this specimen the portals of entry of branches from the abdominal portion of the lumbar artery are shown. These are, in the center, the centrum branches, and above and below vessels derived from the ascending and the descending branches of the lumbar artery. (Reproduced by courtesy of J. B. Lippincott and Company from *Clinical Orthopaedics and Related Research*, No. 115, 1976.)

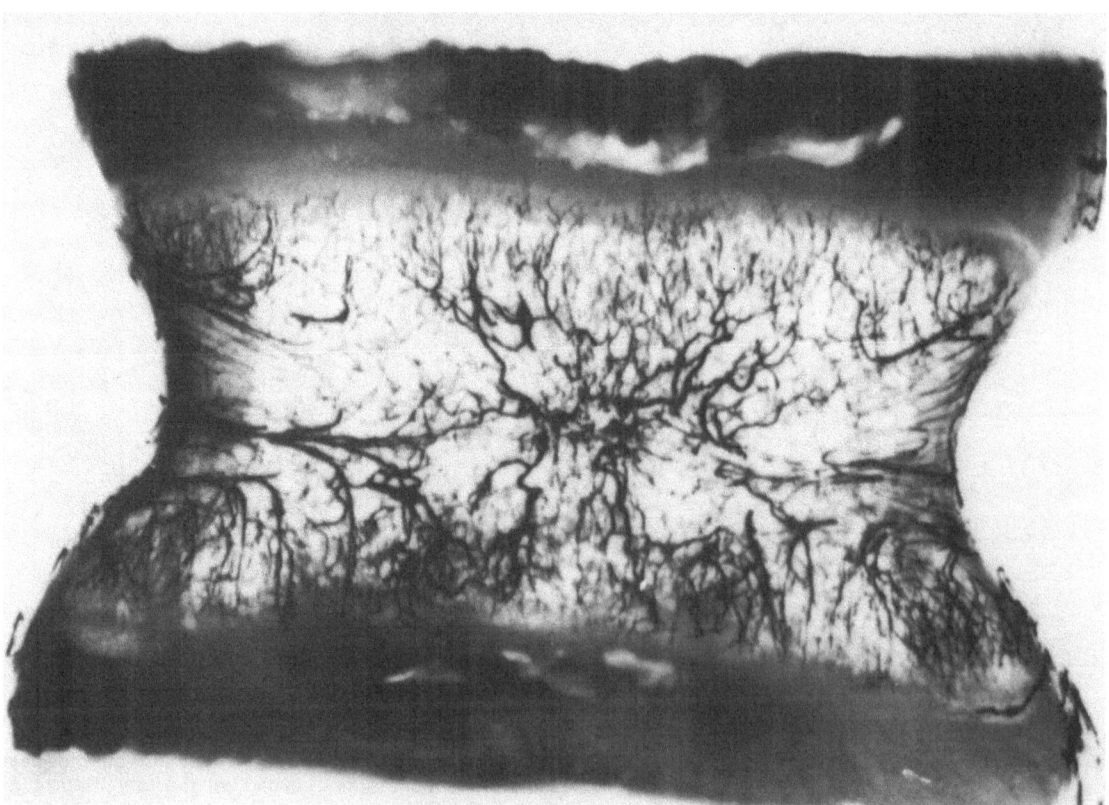

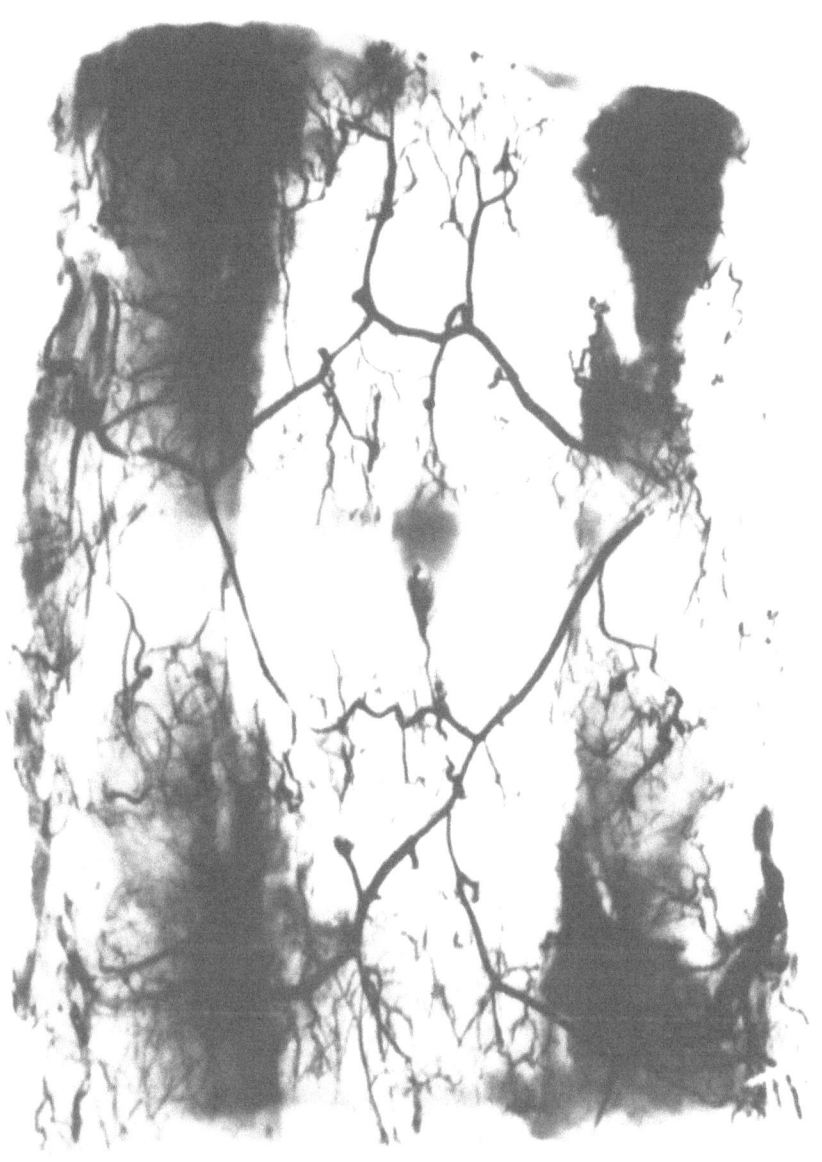

figure 5.4

A radiograph to show the distribution of anterior spinal canal branches
of the lumbar arteries on the posterior surfaces of two adjacent lumbar
vertebral bodies from a male aged 65 years. The main intraosseous
tributaries from this arcuate system correspond to the ascending and
descending branches of the abdominal portion of the lumbar arteries
and to the centrum branches. (Reproduced by courtesy of J. B.
Lippincott and Company from *Clinical Orthopaedics and Related
Research*, No. 115, 1976.)

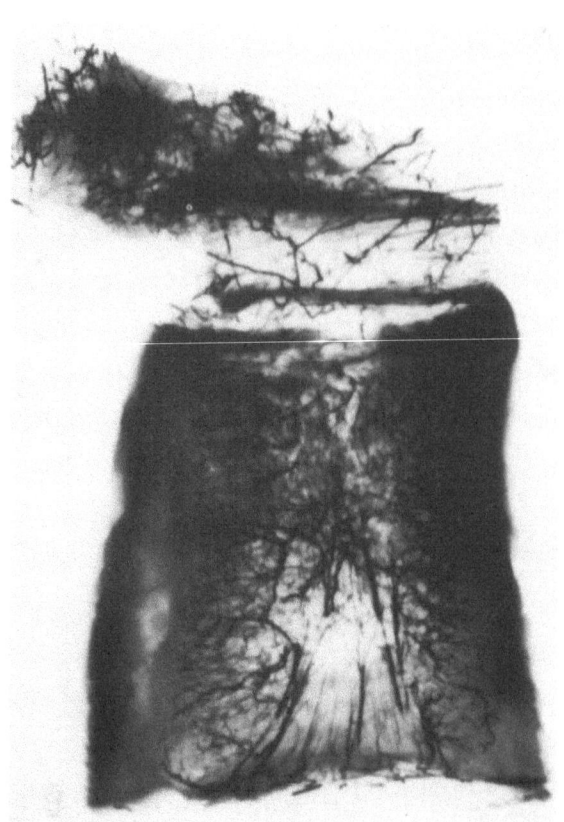

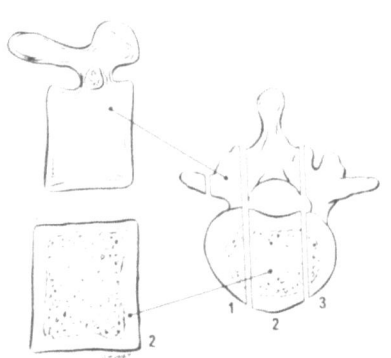

figure 5.5

Line drawings indicating the division
of a lumbar vertebra into sagittal
sections. Sections 1 and 2 of the
specimen alongside show the
distribution of arteries within the
vertebral body from a male aged 30
years. The feeding branches to the
centrum grid can be seen in both
sections. (Reproduced by courtesy
of J. B. Lippincott and Company
from *Clinical Orthopaedics and
Related Research*, No. 115, 1976.)

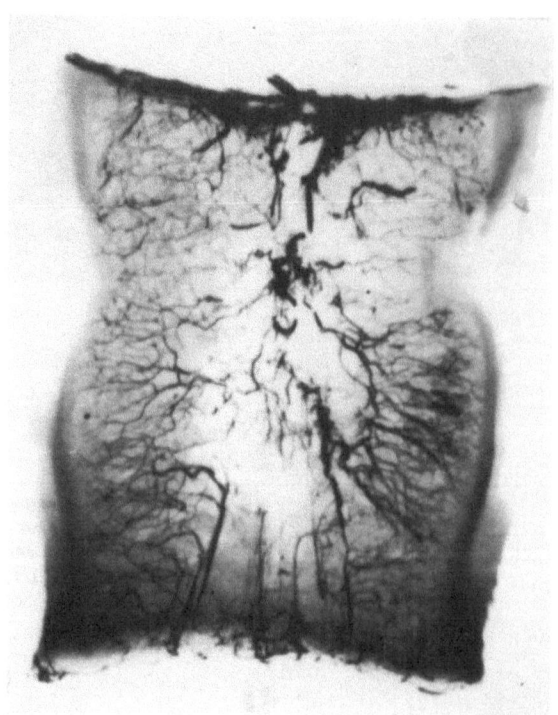

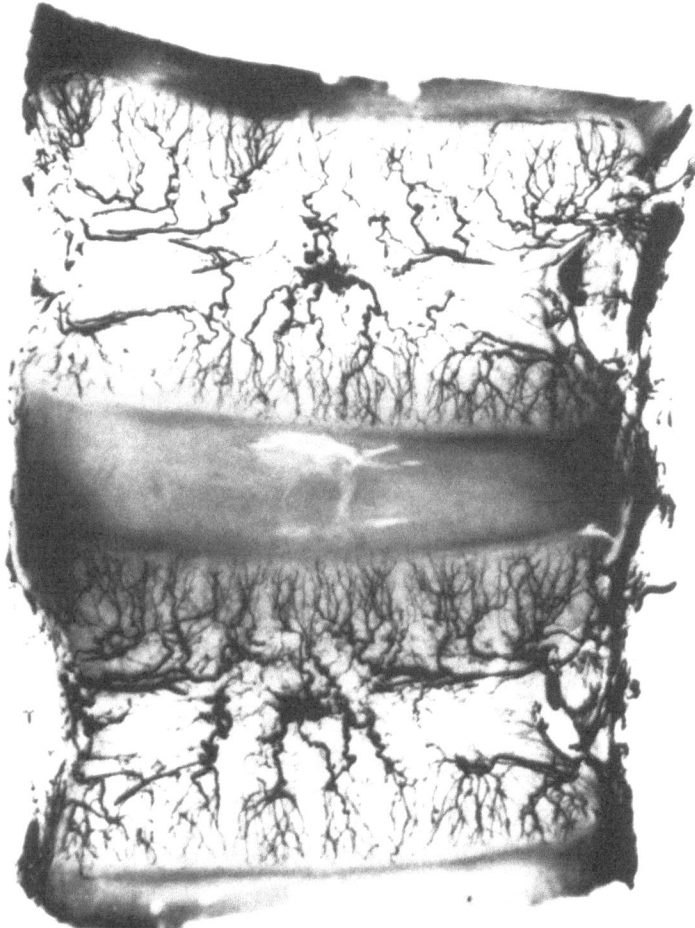

figure 5.6

A thin coronal section through the
center of two adjacent lumbar
vertebral bodies from a male aged
65 years showing details of the
intraosseous distribution of arteries.
(Reproduced by courtesy of J. B.
Lippincott and Company from
*Clinical Orthopaedics and Related
Research*, No. 115, 1976.)

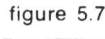

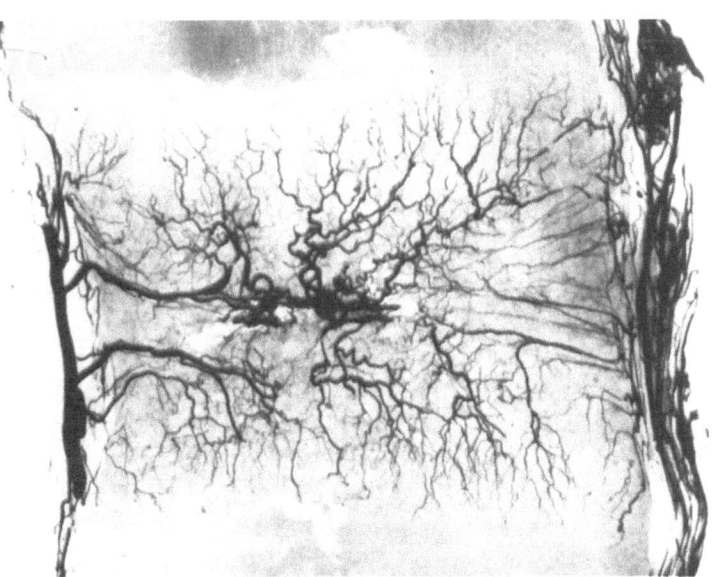

figure 5.7

A thin sagittal section through the
center of the fifth lumbar vertebra
from a young adult showing the
intraosseous arterial patterns
described in the text. (Reproduced
by courtesy of J. B. Lippincott and
Company from *Clinical Orthopaedics
and Related Research*, No. 115,
1976.)

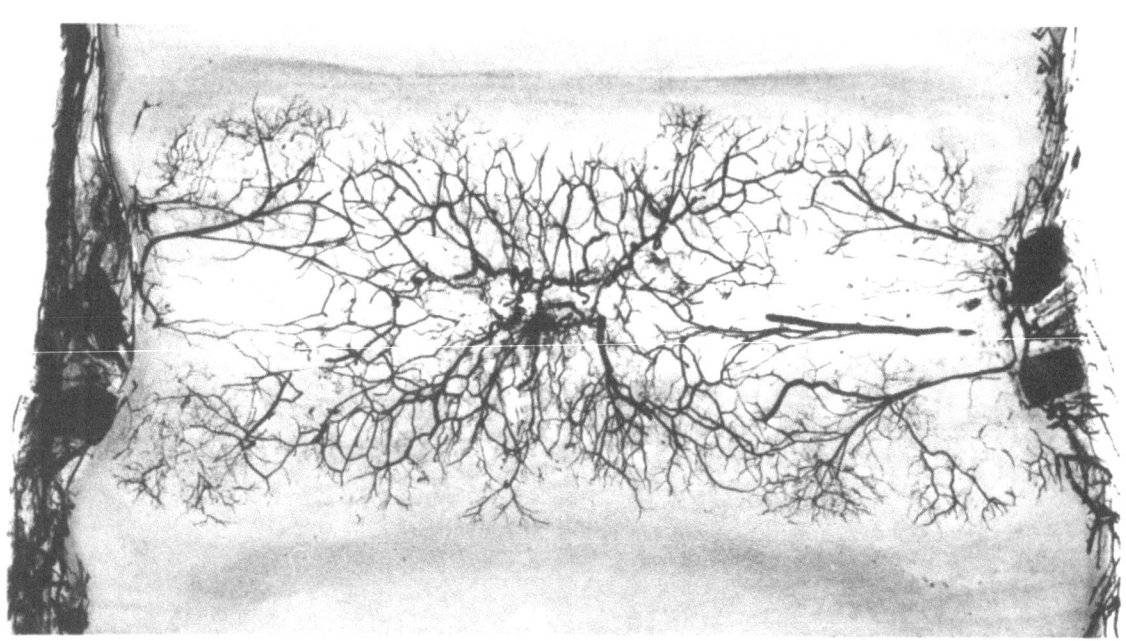

figure 5.8

A thin coronal section from the body of the eleventh thoracic vertebra
of a child aged 13 years. On the right side of the photograph the three
distinct entrant vessels can be seen.

figure 5.9

A radiograph of a thin median
sagittal section of a typical lumbar
vertebra from a newborn infant,
showing the pattern of distribution
of arteries within the vertebral body.
Note the sinusoidal terminations
within the cartilage caps of the
vertebral body.

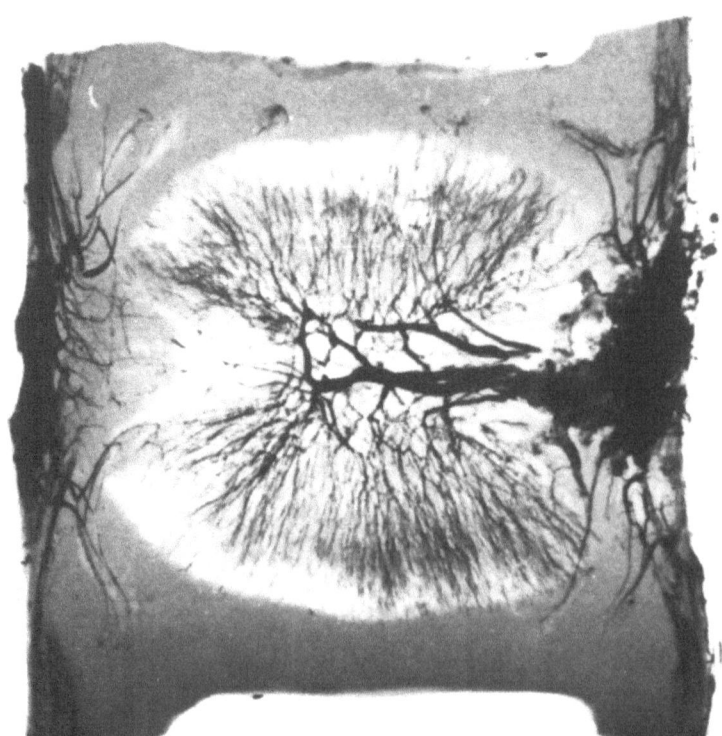

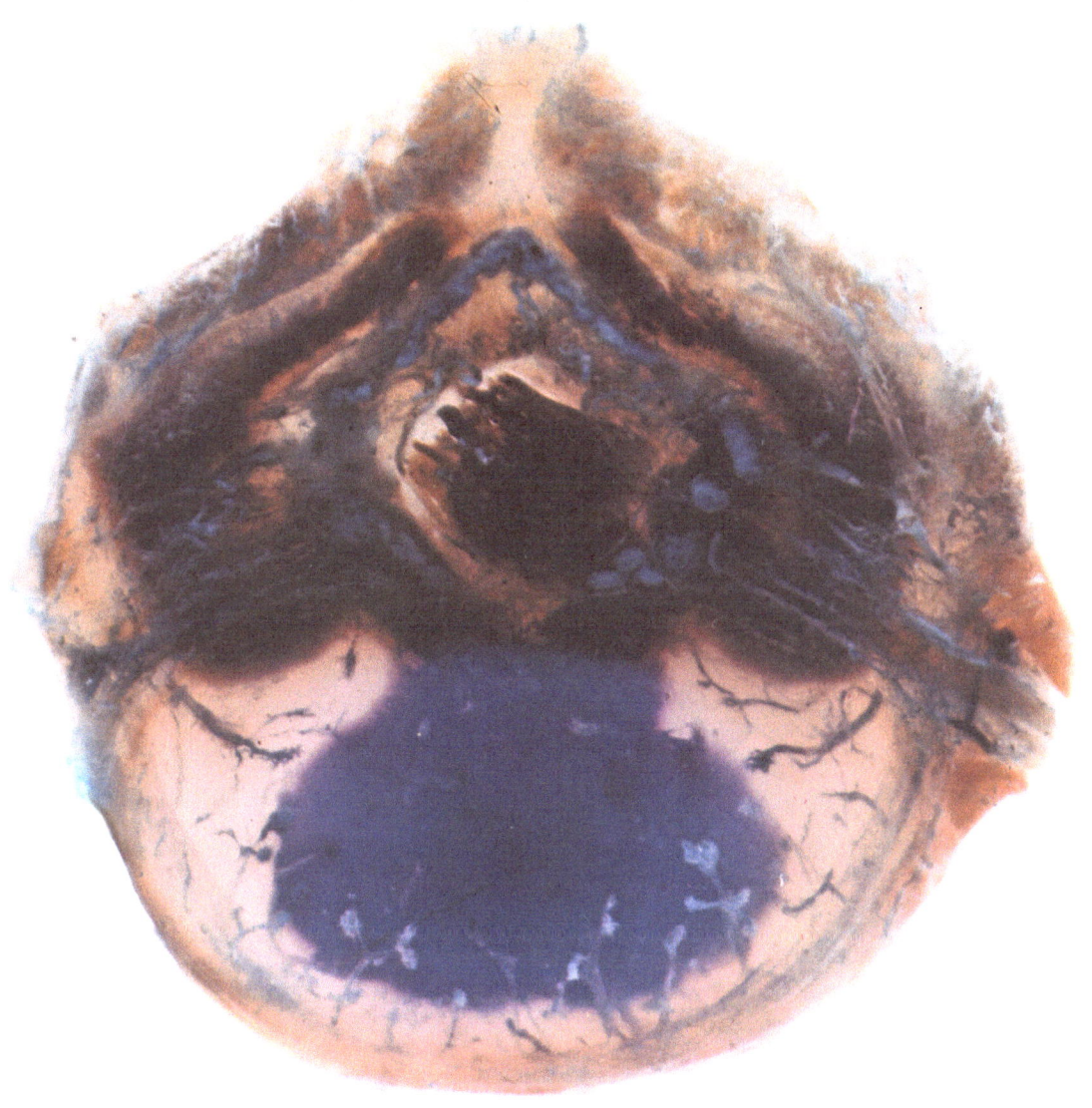

figure 5.10

A photograph of a thin transverse section from a lumbar vertebra in
a newborn child. An arterial injection in which the venous side of the
circulation has also been filled. The specimen is viewed through the
intervertebral disc. The clublike endings of sinusoids in the cartilage
cap of the vertebral body can be seen. The basic radiate pattern of
distribution of the entrant arteries is also shown.

Posteriorly in the spinal canal, the internal vertebral venous plexus is
clearly shown.

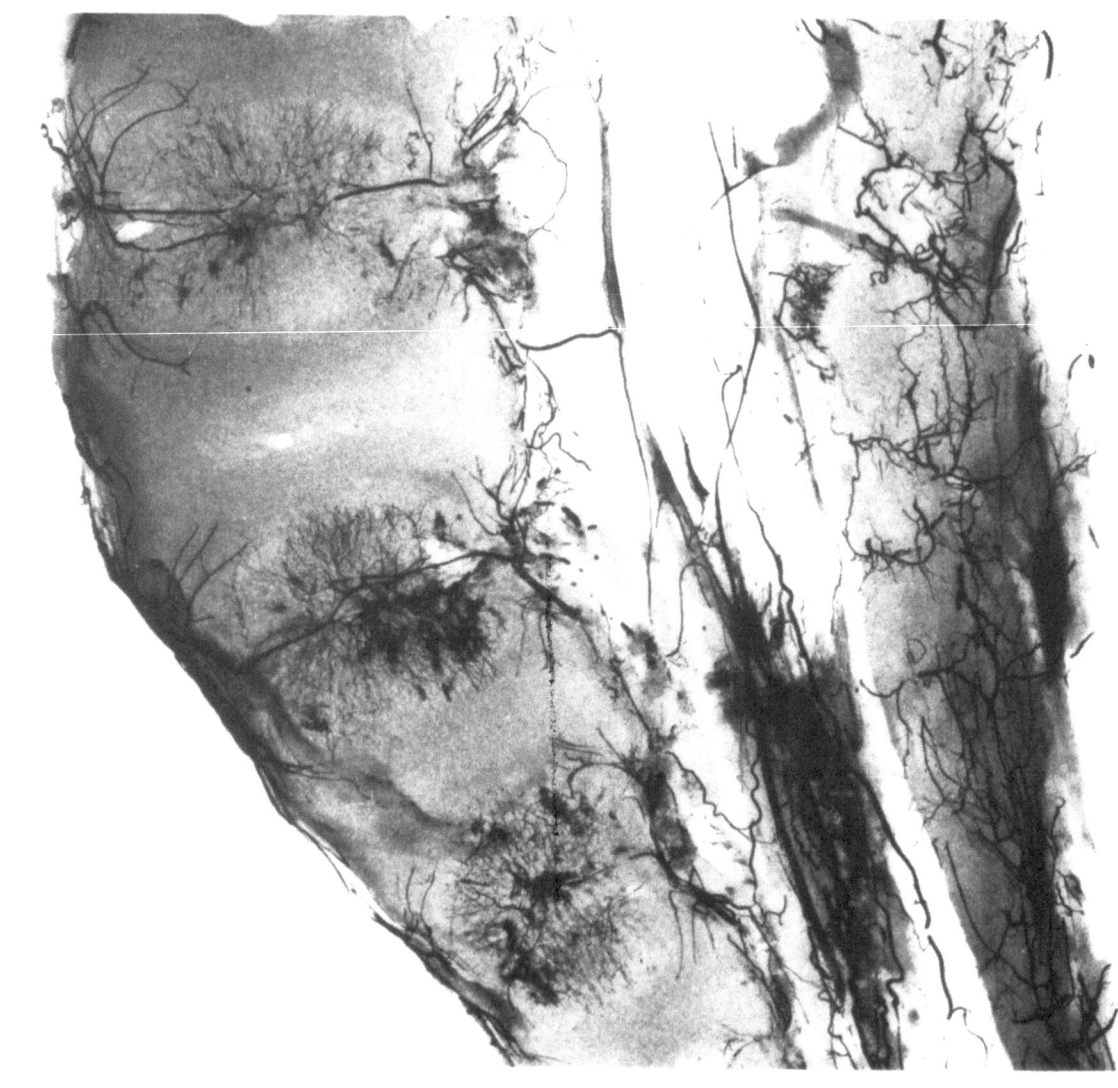

figure 5.11

A photograph of a thin median sagittal section at the lumbosacral
junction from a male child aged 13 months. The cartilage caps on the
ends of the vertebral bodies are clearly shown and the demarcation
between the cartilage end-plate zones and the intervertebral disc
stand out well in this transilluminated specimen. The crude details of
distribution of the arteries within the cartilage can be seen. Note the
Y-shaped bifurcations and clublike endings described in the text.

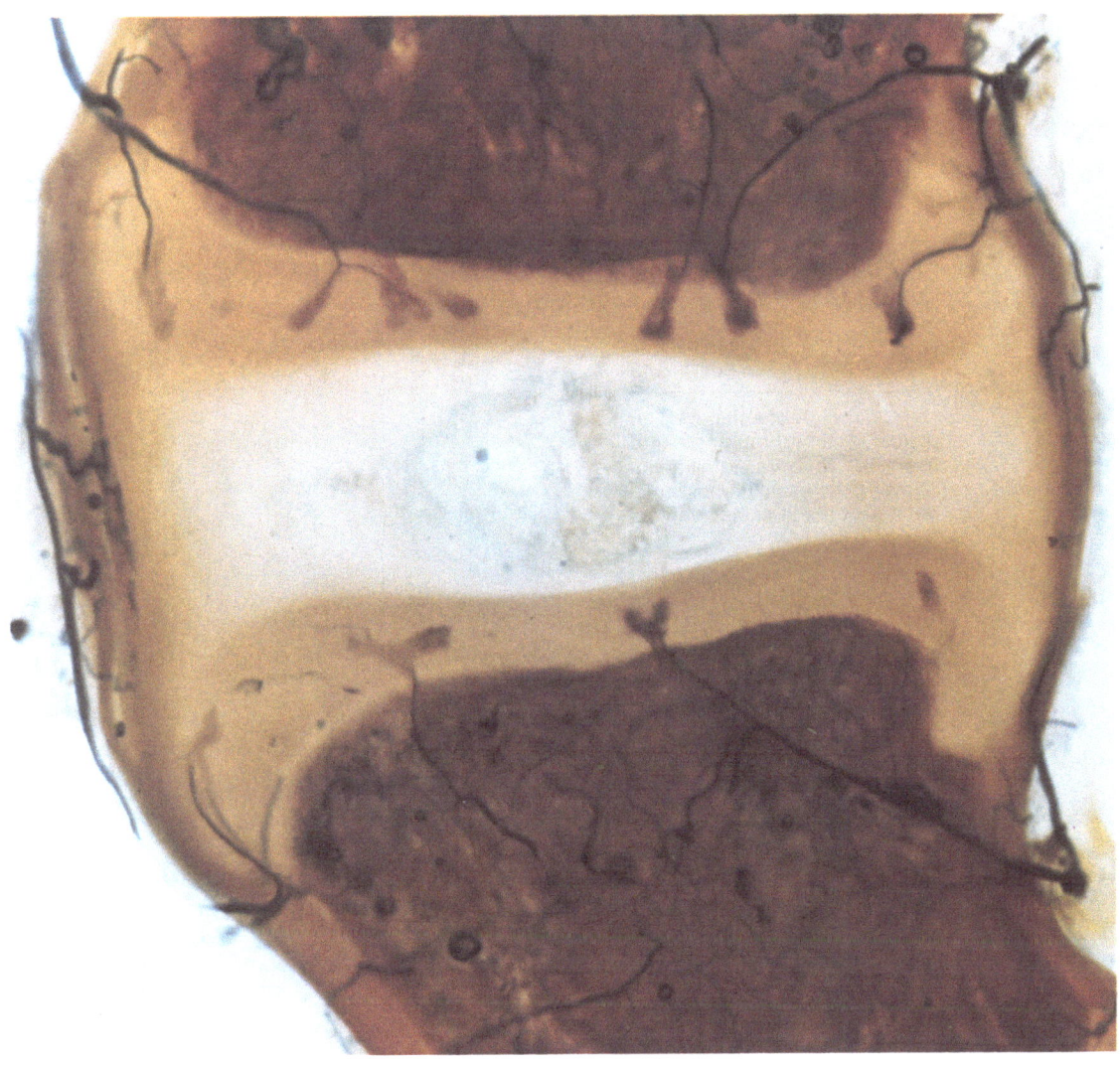

figure 5.12

A radiograph of a thin median sagittal section through the fifth lumbar
and upper two sacral vertebral bodies from a newborn child. The basic
pattern of distribution of the intraosseous arteries is clearly seen. Note
the terminations of some of the vessels in cartilage, particularly those
based on the ascending and descending branches of the lumbar artery
on the anterior surface of the body. Similar branches based on the
anterior spinal canal branches of the lumbar artery can be seen
posteriorly.

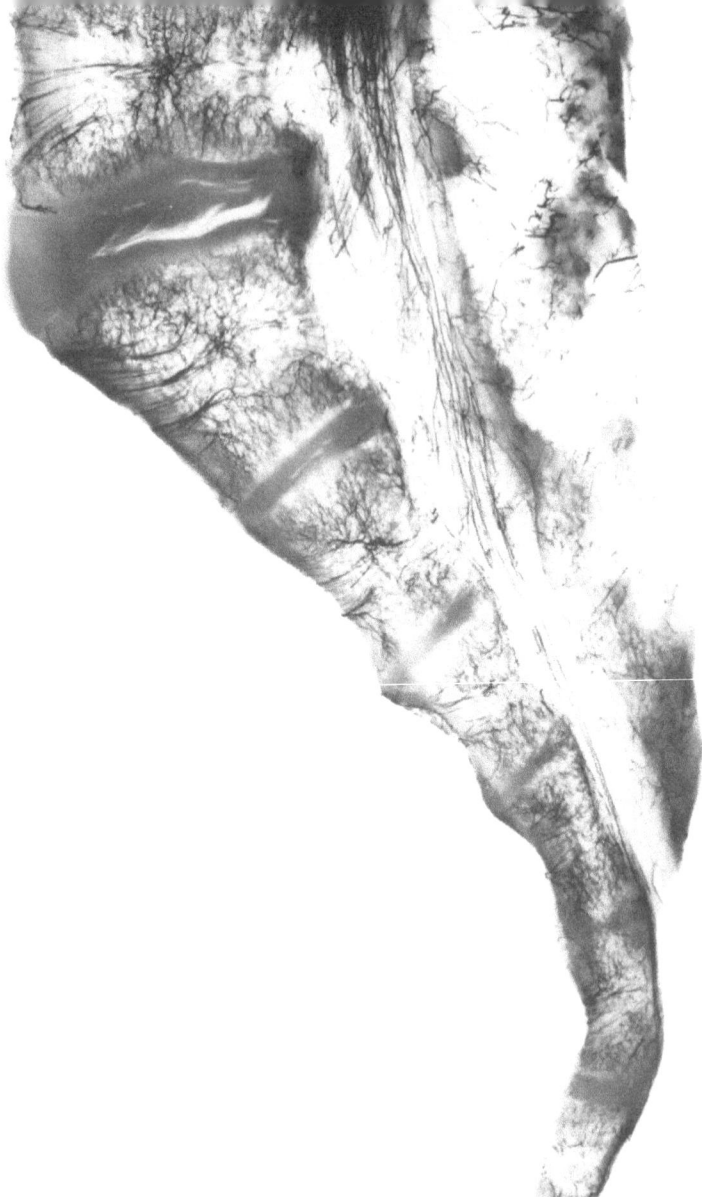

figure 5.13

A radiograph of a thin median
sagittal section through the fifth
lumbar vertebra and sacrum,
showing the basic intraosseous
arterial patterns in an adult spine.

figure 5.14

A thin transverse section to show
details of the arterial distribution
in relation to the spinal canal, the
vertebral body in front, and the
lamina and spinous process behind.
At the level of the intervertebral
foramen, note: 1, the anterior
abdominal wall branches and 2, the
intermediate or spinal canal
branches, and posterior to the
intervertebral foramen, the posterior
branches in relation to the lamina
and spinous process. (Reproduced
by courtesy of J. B. Lippincott and
Company from *Clinical Orthopaedics
and Related Research*, No. 115,
1976.)

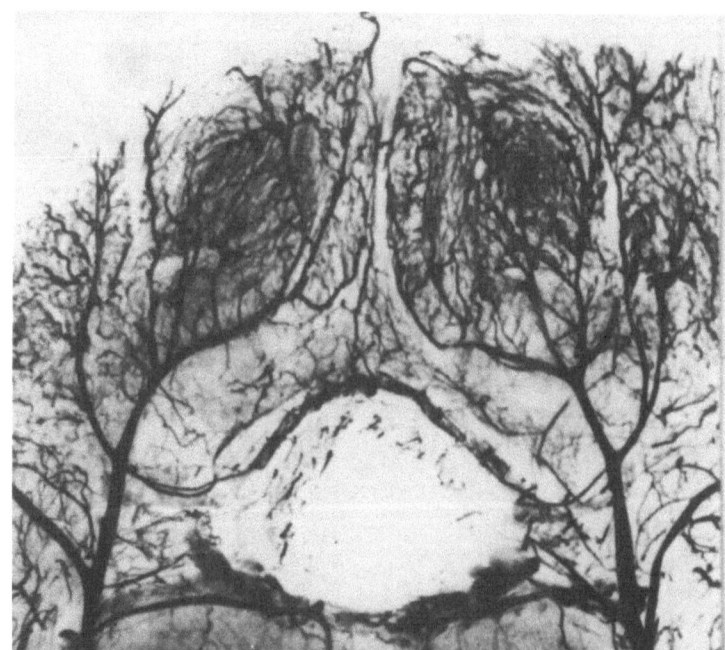

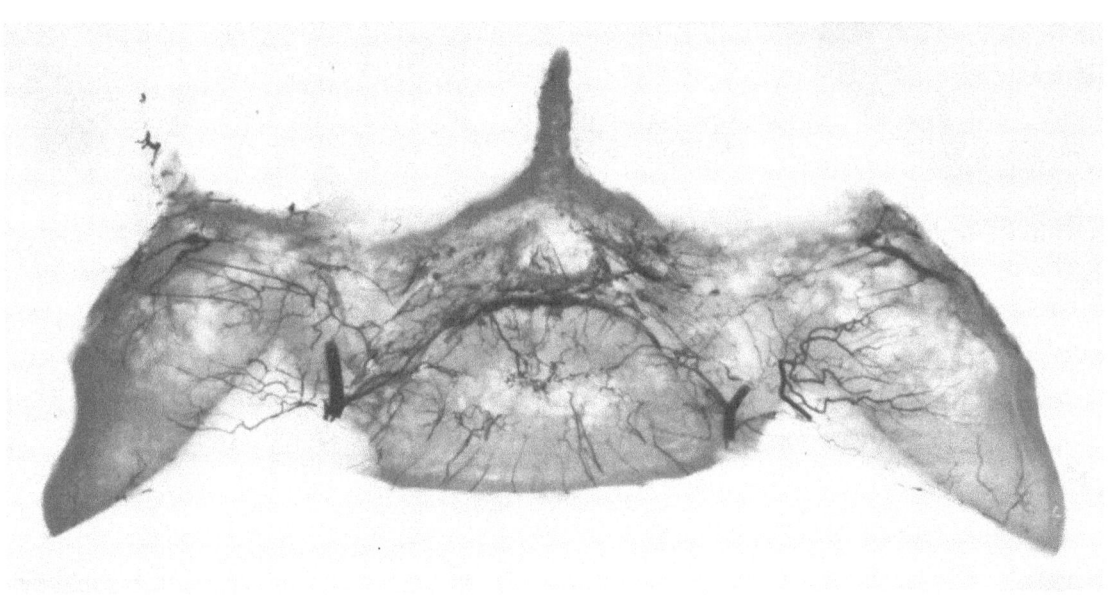

figure 5.15

A radiograph of a transverse section through the alar of the sacrum and
middle of the first sacral vertebral body from an adult, showing the
centrum grid in the vertebral body.

figure 5.16

A radiograph of a midsagittal section
through two adjacent thoracic
vertebral bodies from a male aged
34 years, showing the intraosseous
arteries.

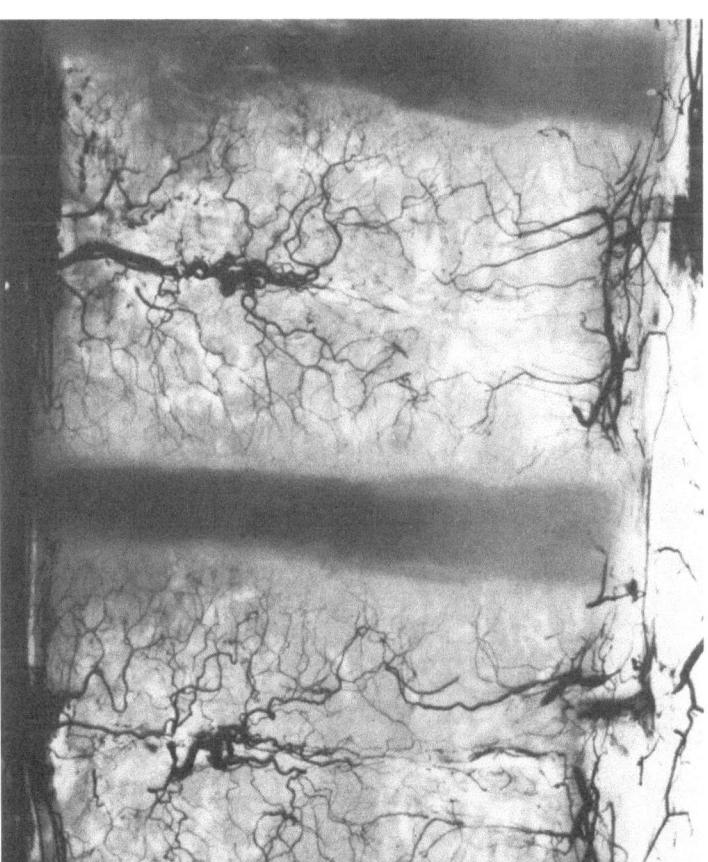

figure 5.17

A transverse section through the vertebral body of a midcervical
vertebra from a female aged 20 years. The section has been cut near
the vertebral end-plate area. On the right side of the photograph,
the origin of the arteries supplying the vertebral body from the vertebral
artery can be seen.

figure 5.18

A radiograph of a thin transverse section through the centrum area
of a typical cervical vertebral body from a female aged 20 years from
the specimen illustrated in Figure 5.17. The radiate vessels contributing
to the centrum grid can be seen. Posteriorly details of the blood
supply of the laminal arch are well shown. The spinal cord is somewhat
distorted, but its intrinsic arterial circulation is well shown.

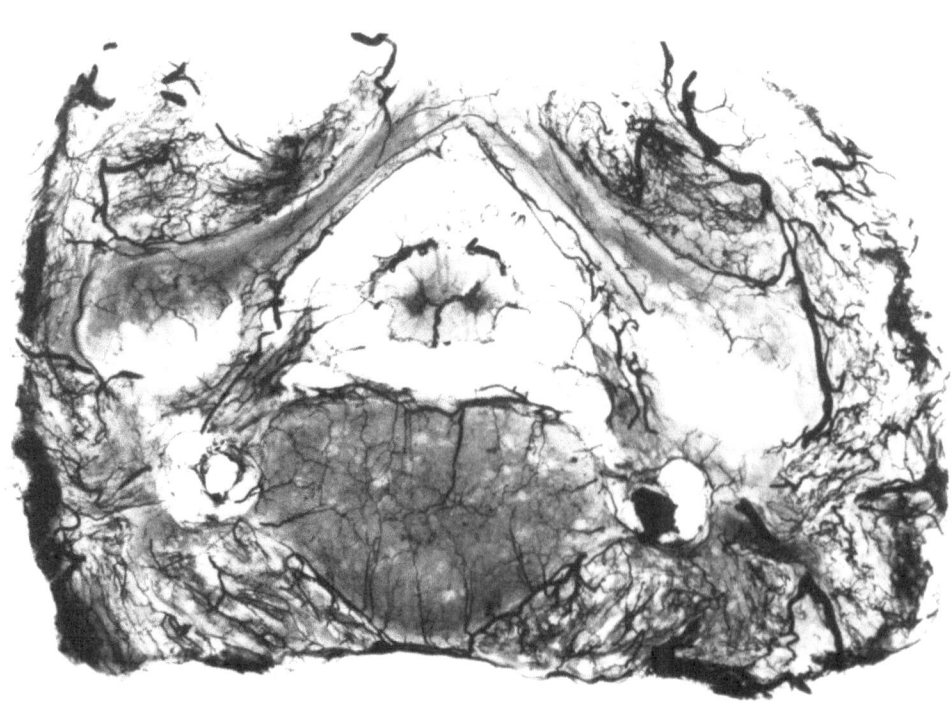

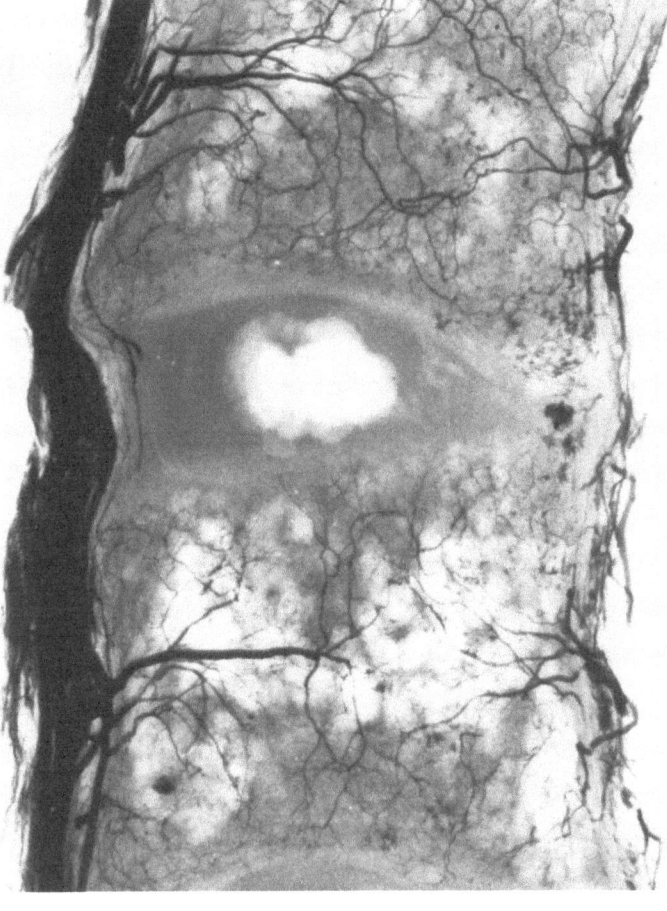

figure 5.19

A radiograph of a thin median
sagittal section through two typical
adult cervical vertebral bodies from a
female aged 20 years, showing the
pattern of distribution of arteries
within the vertebrae.

figure 5.20

A radiograph showing the distribution of arteries to the facet joints of
the atlas from an adult.

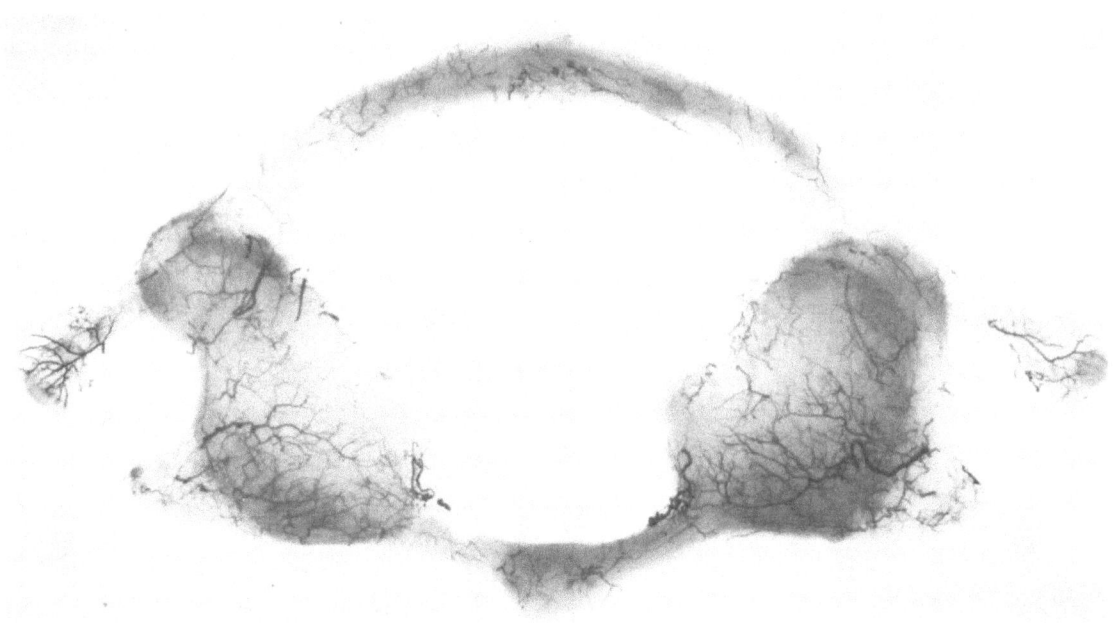

figure 5.21

A transverse section through the atlas from the spine of a female aged
20 years with the odontoid process of the axis shown in the plane of
section showing the arterial supply of the odontoid peg and of the
facet joints of C2.

figure 5.22

A transverse section through the region of the center of the body of
an upper thoracic vertebra from the body of a male child aged 9
months. The neurocentral joints are shown on both sides with radiate
arteries in the cartilages terminating in sinusoidal systems with bulbous
ends.

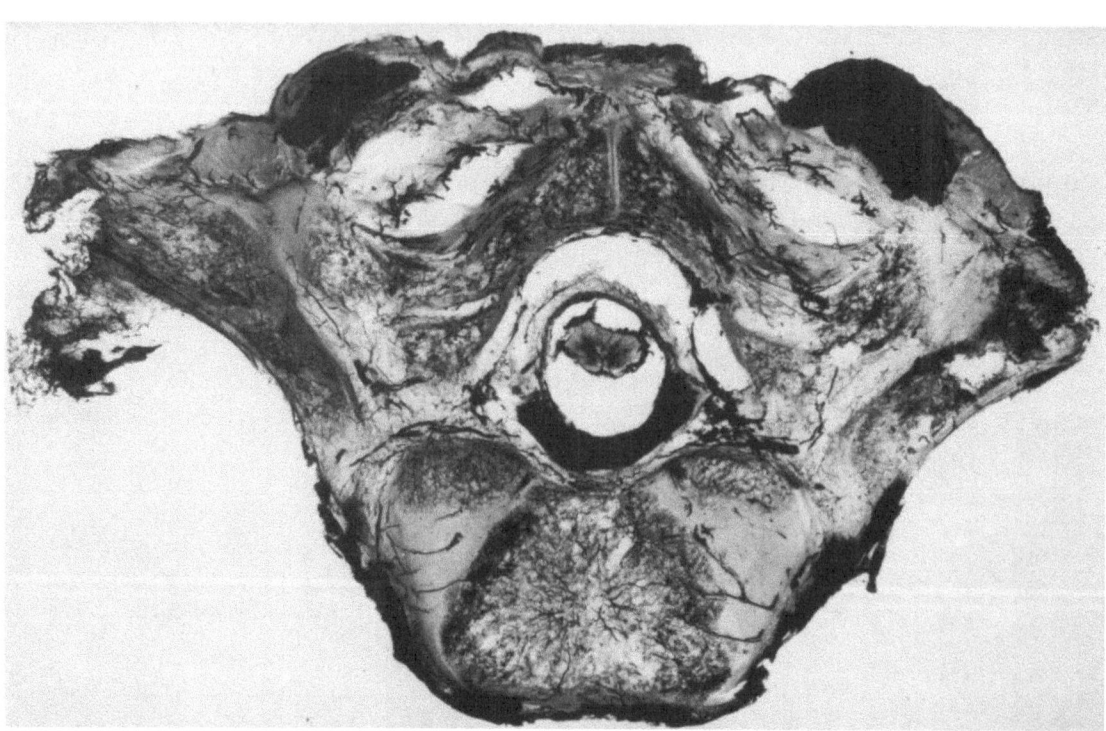

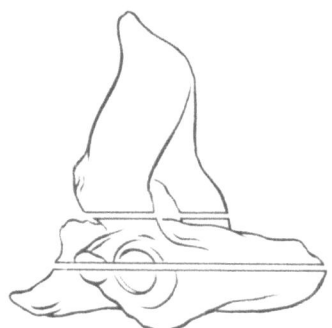

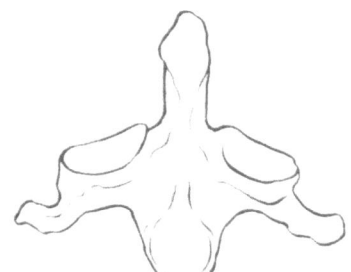

figure 5.23

A line drawing indicating the coronal planes of sectioning of a second
cervical vertebra, with the outline of the anterior segment alongside.
The radiograph shows the distribution of arteries within the bone.

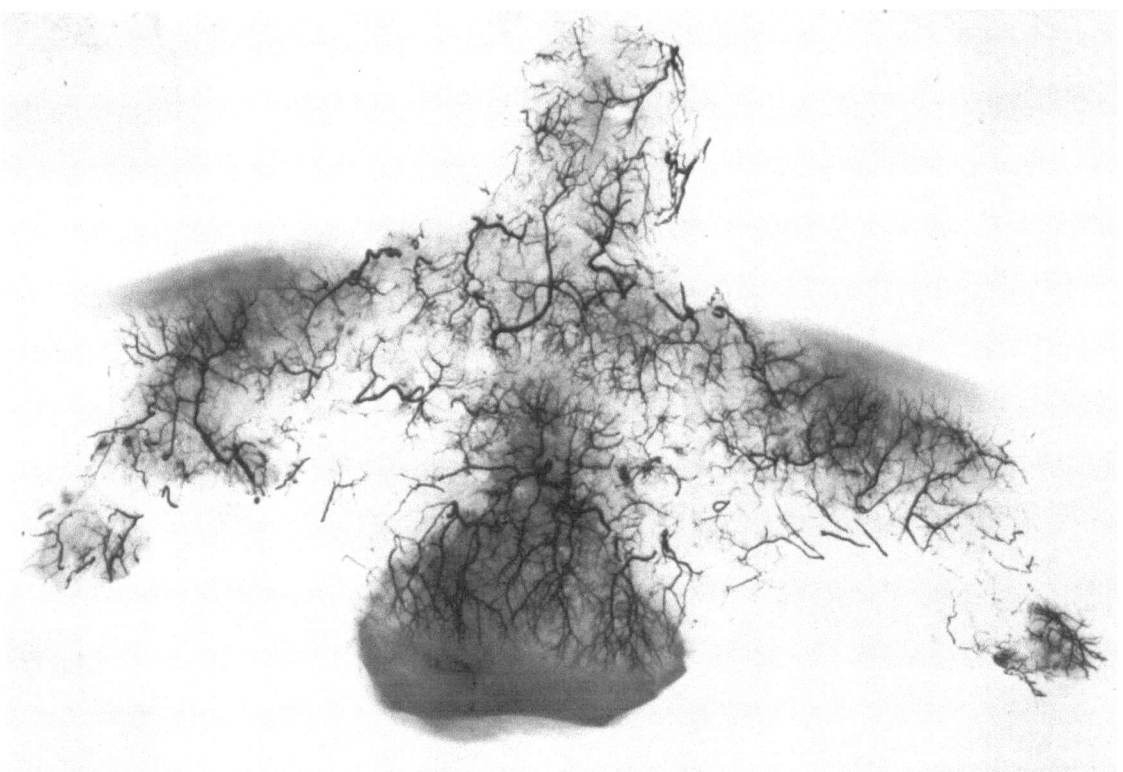

6
Veins
of the
vertebral body

The basivertebral system of veins is orientated horizontally in the centrum. It is arranged in the middle of the vertebral body along with the radiate arteries, forming a large-scale venous grid into which the vertical veins of the vertebral body flow from above and below. The basivertebral veins converge posteriorly to drain into the anterior internal vertebral venous plexus, sometimes as a single vein, sometimes as two separate tributaries. Anteriorly they join the external vertebral venous plexus.

The main vertical venous channels are of large caliber and run gently tortuous courses. They are formed by the confluence of numbers of equally large branches which enter the main stems obliquely and at regular intervals along their courses and around the circumferences. Individual branches themselves are formed by the union of innumerable short, fine radicles. The patterns of the intraosseous arteries in the vertebral bodies may be described graphically as resembling the fine branches of a deciduous tree, such as a cherry tree which has shed its leaves. Intraosseous veins more closely resemble the fibrous root system of common plants, in which many of the roots are of similar size with clusters of fine radicles called root hairs at their ends.

In the region of the vertebral body adjacent to the vertebral end-plate, large venous channels are found orientated horizontally, running parallel to the end-plate area when viewed in sagittal or coronal sections. This venous system is analogous to the subarticular collecting vein system in the juxtaarticular zones of the bones of the lower limb. In the vertebral body we have described this as *the horizontal subarticular collecting vein system* (Crock, Yoshizawa, and Kame, 1973). It is built up in the central area of the vertebral body by large caliber tributaries of the vertical veins of the centrum, which turn abruptly from their vertical courses to run horizontally, some passing anteriorly, others posteriorly, and still others laterally. In the posterior part of the vertebral body some of the tributaries from this horizontally oriented network run directly into the anterior internal vertebral venous plexus. Anteriorly and around the circumference of the vertebral body, tributaries of veins draining directly into the external vertebral venous plexus also contribute to the formation of the horizontal subarticular collecting vein system.

At the vertebral end-plate level we have found another vascular network of smaller caliber that is orientated horizontally and parallel to the subarticular collecting vein system. This lies on the perforated cortical vertebral end-plate, forming what we have named *the subchondral postcapillary venous network of the vertebral body*. Short, vertical tributaries from this network drain into the horizontal subarticular collecting vein system, while peripherally some tributaries drain directly into adjacent veins on the surface of the vertebral body.

We believe that this subchondral postcapillary venous network receives tributaries at right angles to its plane of orientation from capillaries in the vertebral end-plate cartilage, that is, from *the vertebral end-plate cartilage capillary bed*.

Studying thin sections from the vertebral bodies in newborn infants in which the venous system has been filled, it will be found that the main channels are orientated in the same way as the entrant arteries which are described in detail in Chapter 5. These basic patterns are found throughout life, again with the exception that in the adult the complex subarticular collecting vein system is formed.

The spinous process is drained by a central vein which connects the external vertebral venous tributaries with the posterior internal vertebral venous plexus. Similarly in the lamina large venous tributaries course toward the pedicles, joining with veins from the pedicles to emerge in the region of the intervertebral foramina.

We have attempted to illustrate all the features described in the accompanying figures.

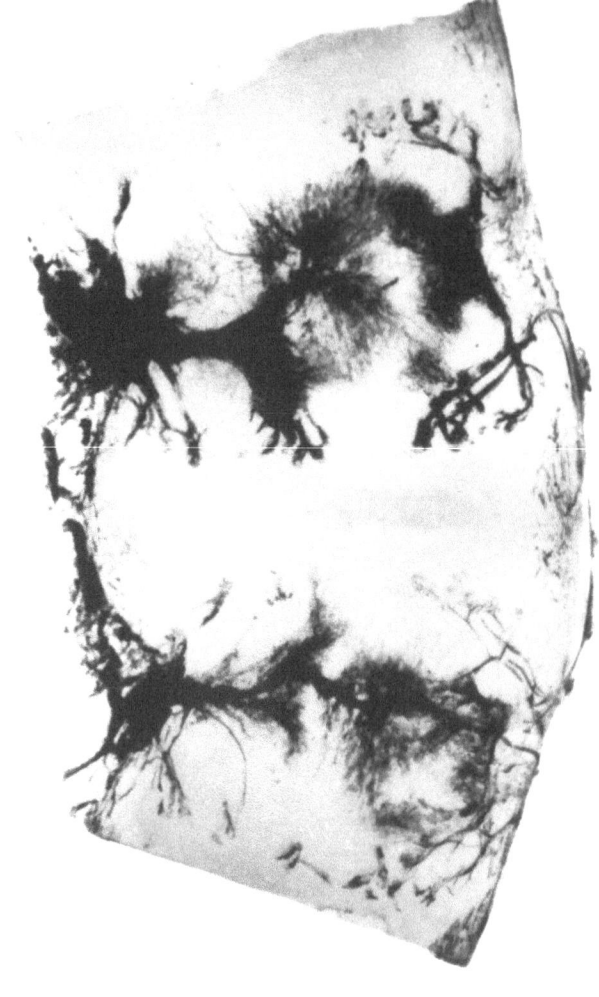

figure 6.1

A radiograph of a thin median
sagittal section at the lumbosacral
junction in a newborn child. The
arterial system has been injected,
and almost complete venous filling
achieved also. The major outlines of
the basivertebral central grid system
are shown in both vertebrae. Venous
branches corresponding to the
entrant arteries are also shown and
the clublike terminations of the
intracartilaginous sinusoidal vessels
can be seen.

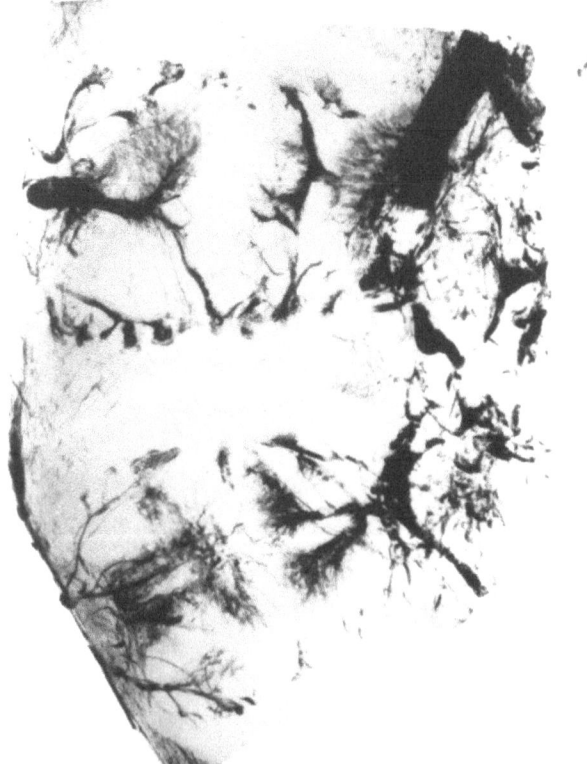

figure 6.2

A thin sagittal section from the same
specimen cut near the pedicles
laterally, showing good venous filling
in the body of L5 and arterial filling
in the body of S1 with partial venous
filling posteriorly of the basivertebral
grid system.

The blood supply of the vertebral column
and spinal cord

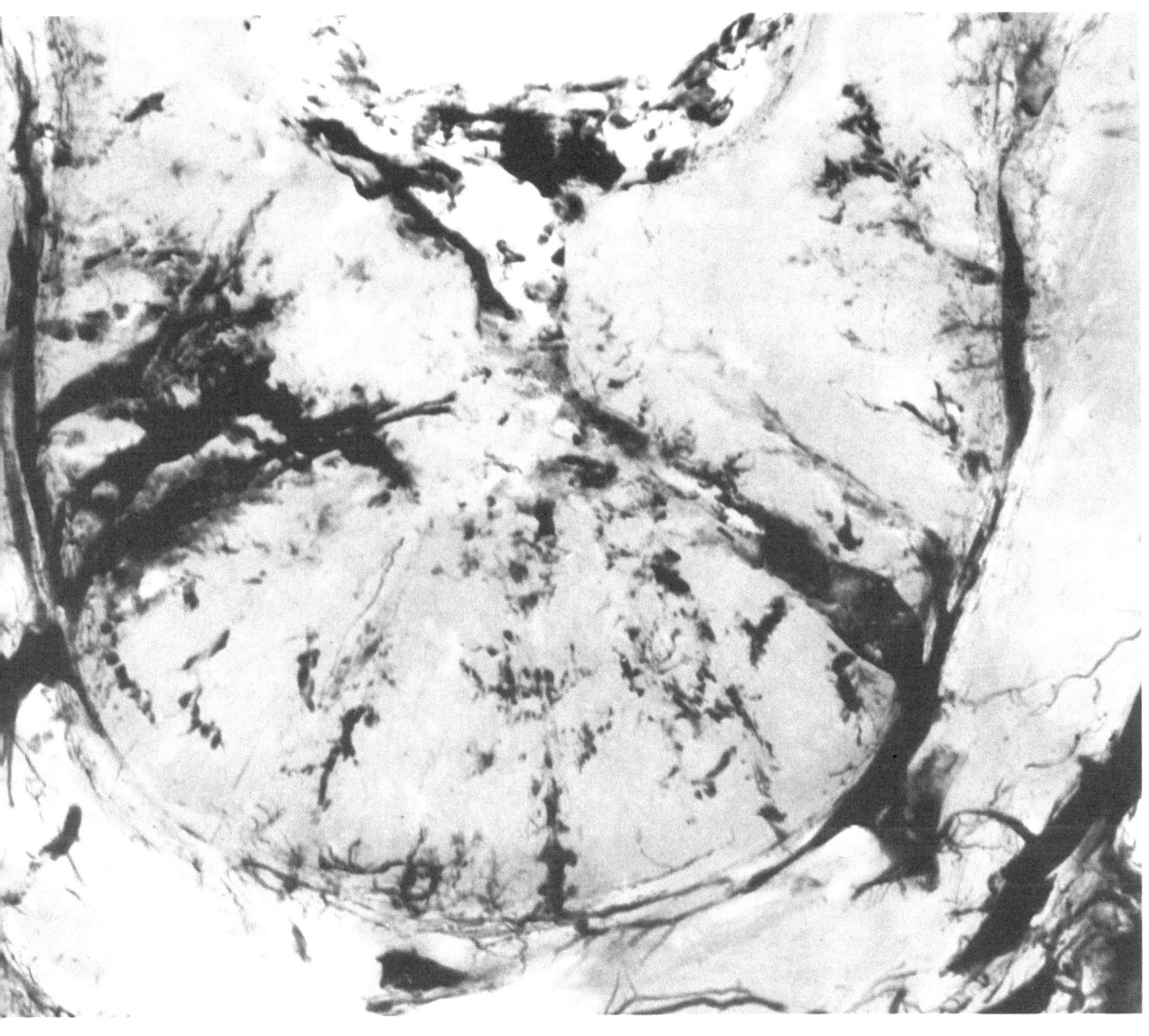

figure 6.3

A radiograph of a thin transverse section from the center of a lumbar
vertebral body from a man aged 29 years. The basivertebral veins are
partly filled with Micropaque. The radiate pattern of the main branches
is shown. Posteriorly the system drains by a single large stem into the
anterior internal vertebral venous plexus.

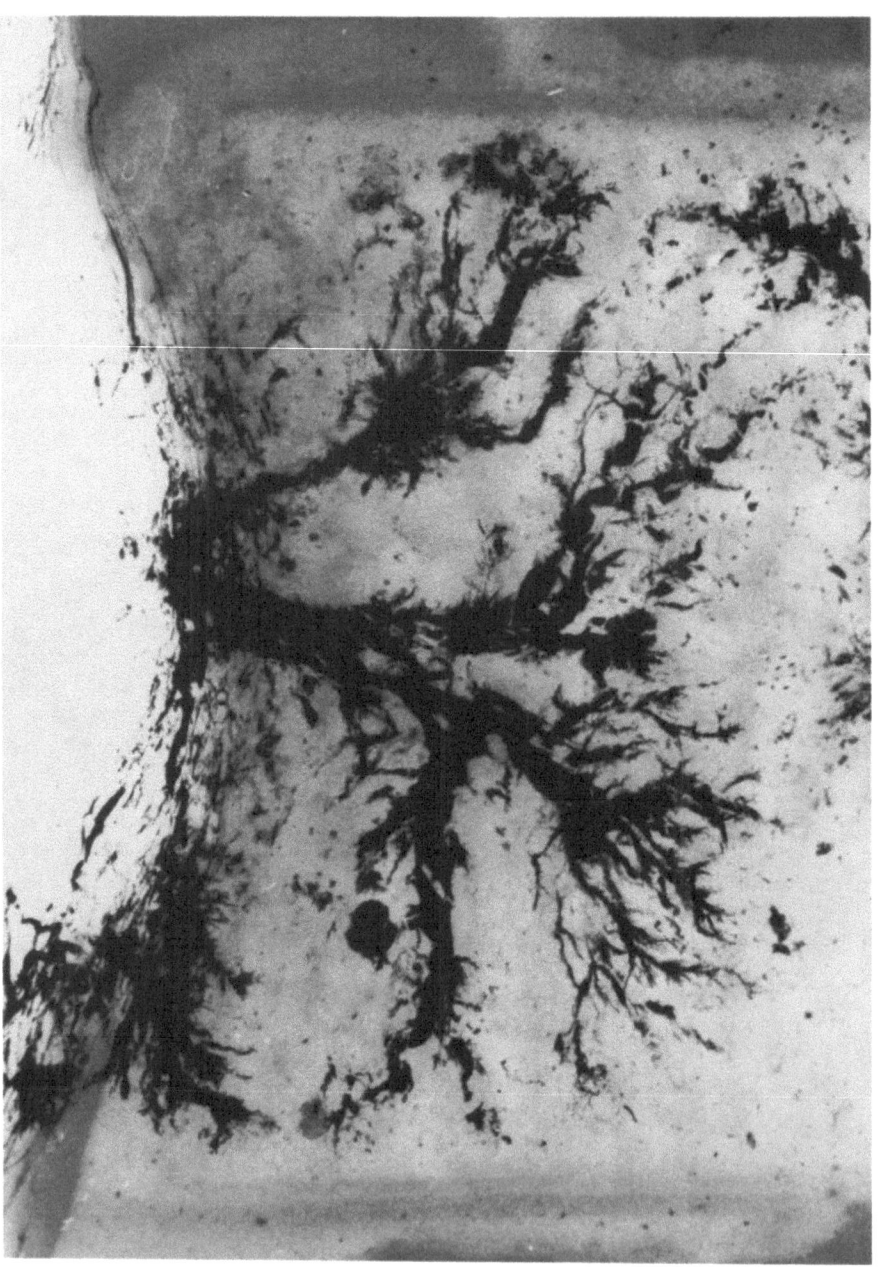

figure 6.4

A radiograph of a thin coronal section of a lumbar vertebral body
from a male aged 60 years. Details of the basivertebral grid system can
be seen in the center of the body. In addition, the venous radicles
which correspond to the entrant arteries based on the ascending and
descending branches of the lumbar artery can be seen partly filled.

figure 6.5

A radiograph of a thin coronal section near the central area of a lumbar
vertebra from a man of 60 years. Note the stellate arrangement of
tributaries draining into the central vein of the basivertebral grid
system. These particular vessels almost certainly drain the marrow
spaces of the vertebral body. The longer vertical tributaries relate to
the complex vertebral end-plate drainage system.

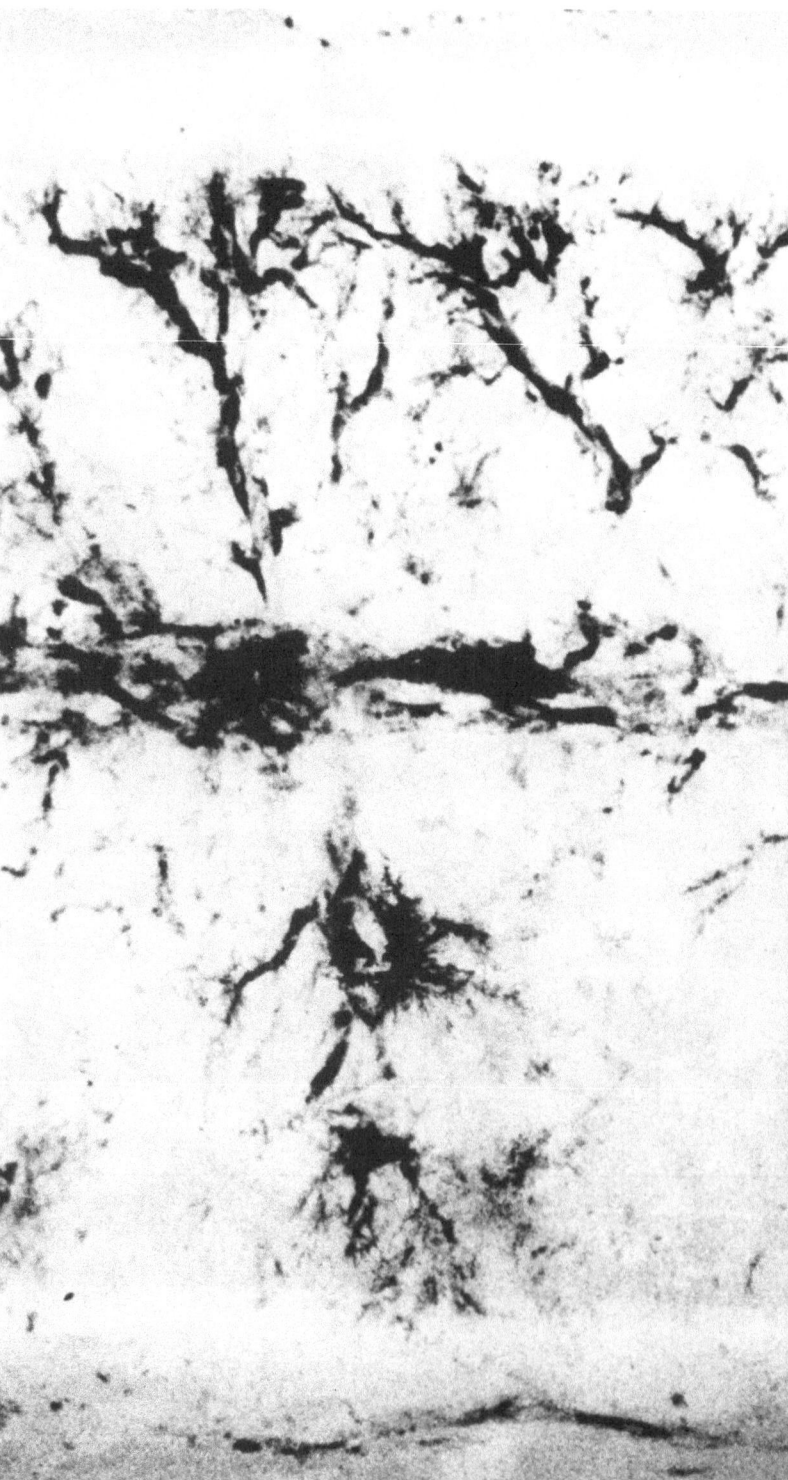

figure 6.6

A radiograph of a thin central
coronal section of a thoracic
vertebral body of a male aged 80
years. Details of the descending and
ascending vertical veins of the
vertebral body are shown. The veins
are partly filled with Micropaque and
the specimen gives an indication of
the magnitude of the vertical
tributaries of the basivertebral grid
system. Venous filling in the region
of both vertebral end-plates is
incomplete. (Reproduced by
courtesy of J. B. Lippincott and
Company from *Clinical Orthopaedics
and Related Research*, No. 115,
1976.)

figure 6.7

A radiograph of a thin sagittal
section from the center of the fifth
lumbar vertebral body of a male
aged 72 years. Again the basivertebral
grid can be seen partly filled with
Micropaque as are the main stems of
ascending and descending vertical
veins of the vertebral body. It is
important to note that venous filling
is confined in this case only to the
largest venous radicles in the
vertebral body. Nonetheless, it is a
valuable specimen illustrating some
of the basic features of the anatomy
of the basivertebral grid system
relevant to its vertical tributaries
and to the anatomy of some of their
own lateral tributaries. These veins
are of large caliber and during
preparation much of the injected
mass often falls out of them.

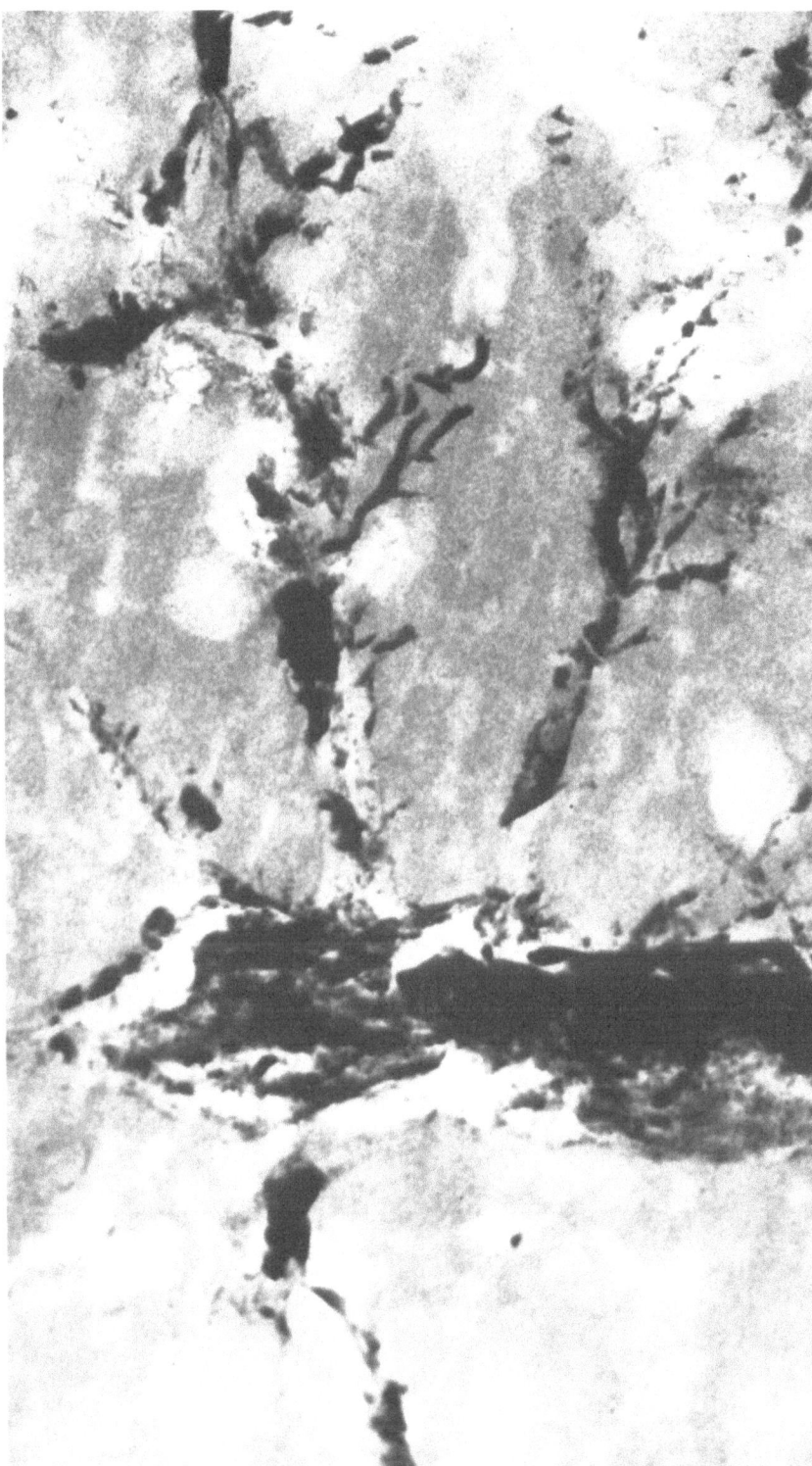

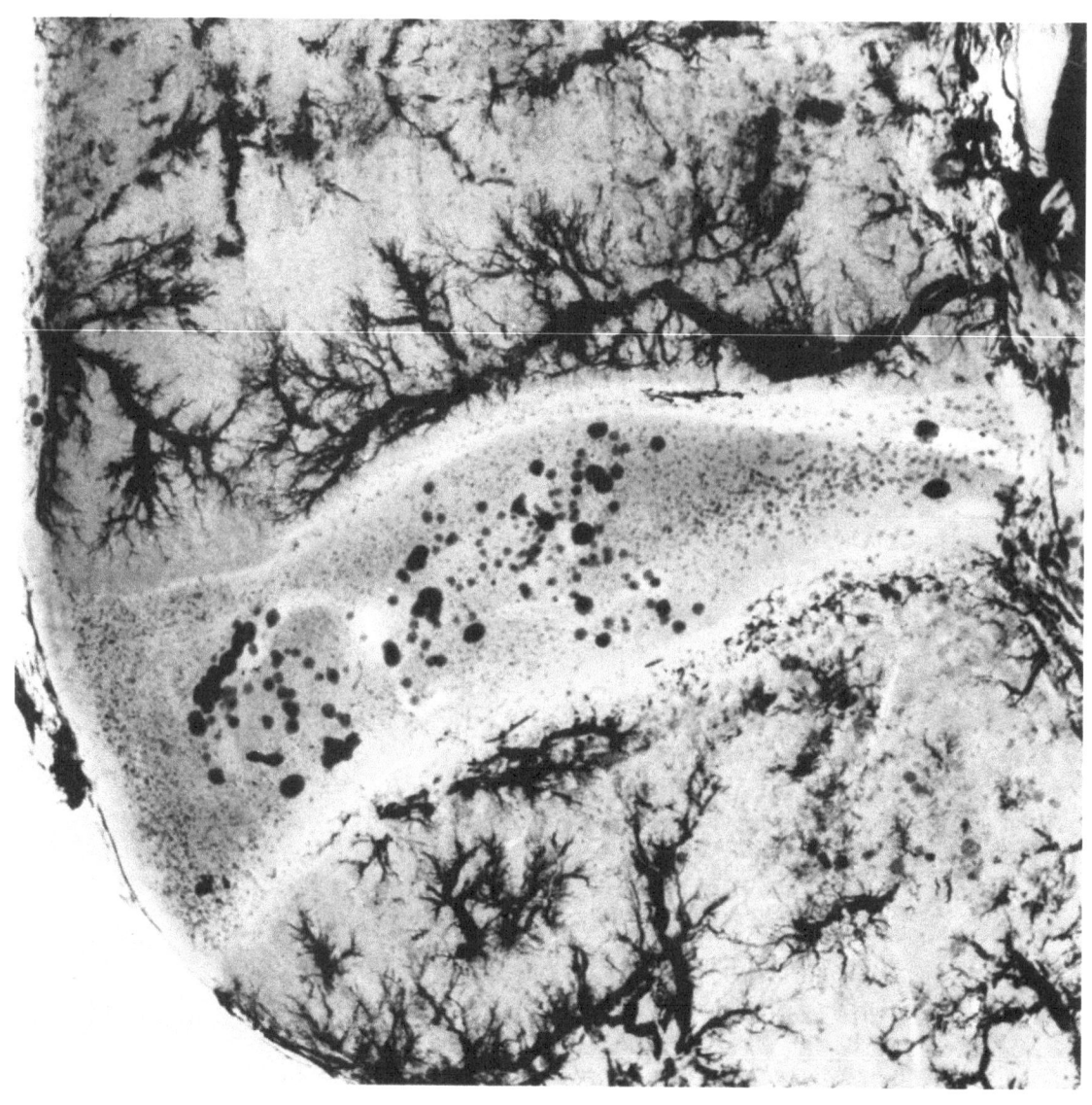

figure 6.8

A radiograph of a thin sagittal section cut laterally near the vertebral pedicle from the lumbosacral junction of a woman aged 67 years. Some fragments of barium sulfate debris have adhered to the cut surface of the disc. The horizontal subarticular collecting vein system of the vertebral body can be seen running parallel to the vertebral end-plate area on the lower surface of the fifth lumbar vertebra. Nearer the disc, of smaller caliber, running parallel to the vertebral end-plate cartilage, the subchondral postcapillary venous network can be seen. It is only partly filled. This system drains by vertical stems through perforations in the vertebral end-plate into the larger horizontal subarticular collecting vein system. In this specimen only one such stem can be seen joining these two venous channels. (Reproduced by courtesy of the editor, *J. Bone Joint Surg.* 55-B, 1973 and J. B. Lippincott and Company from *Clinical Orthopaedics and Related Research*, no. 115, 1976.)

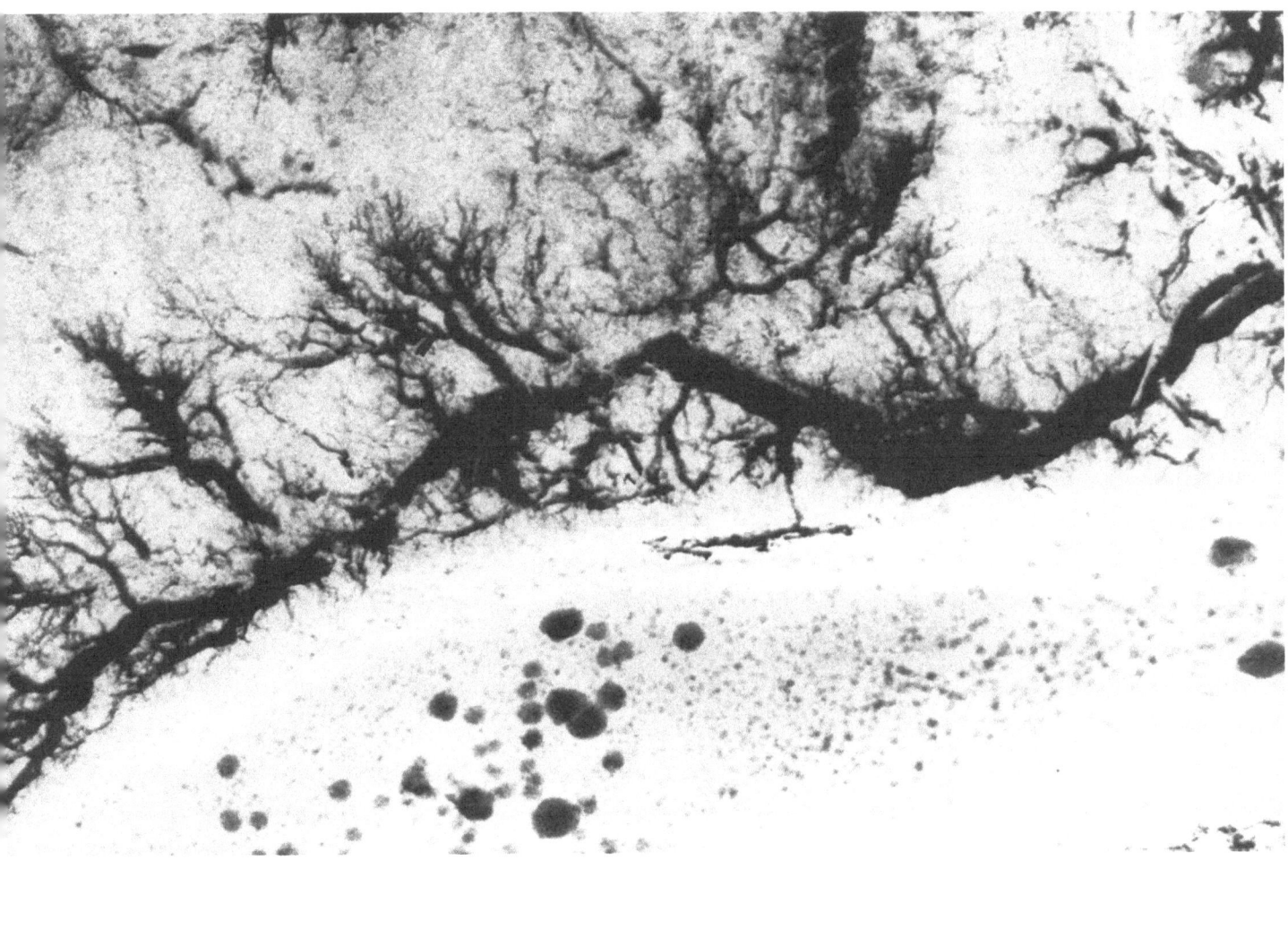

figure 6.9

A detail from the radiograph illustrated in Figure 6.8, to show the
subchondral postcapillary venous network in greater detail.
(Reproduced by courtesy of the editor, *J. Bone Joint Surg.* 55-B, 1973,
and J. B. Lippincott and Company from *Clinical Orthopaedics and
Related Research,* No. 115, 1976.)

figure 6.10

A radiograph of a thin sagittal section from the center of a cervical vertebra from a man of 60 years, showing a detailed view of *the subchondral postcapillary venous network.* Part of a vertical tributary of the basivertebral vein system is seen in the center of the field. The horizontal subarticular collecting vein system is not filled in this specimen.

figure 6.11

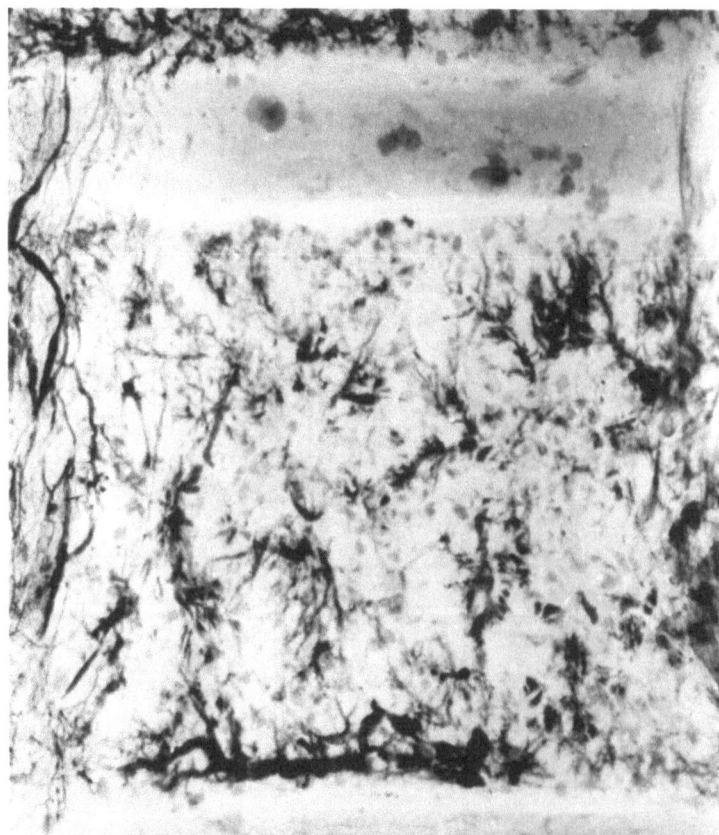

A radiograph of a thin coronal section from a thoracic vertebra, from a woman of 67 years. The horizontal subarticular collecting vein system can be seen in the central third of the field, adjacent and parallel to the vertebral end-plate area. (Reproduced by courtesy of the editor, *J. Bone Joint Surg.* 55-B, 1973, and J. B. Lippincott and Company from *Clinical Orthopaedics and Related Research,* No. 115, 1976.)

figure 6.12

A detailed photograph taken from the central area of the disc and
vertebral body (500 μm thick), from a woman of 30 years. Spalteholz
cleared specimen, approximately ×20. The demarcation line between
the intervertebral disc and vertebral end-plate cartilage is clearly
visible. *The vertebral end-plate cartilage capillary bed* is shown, with
vertical tributaries draining to the subchondral postcapillary venous
network orientated parallel to the vertebral end-plate.

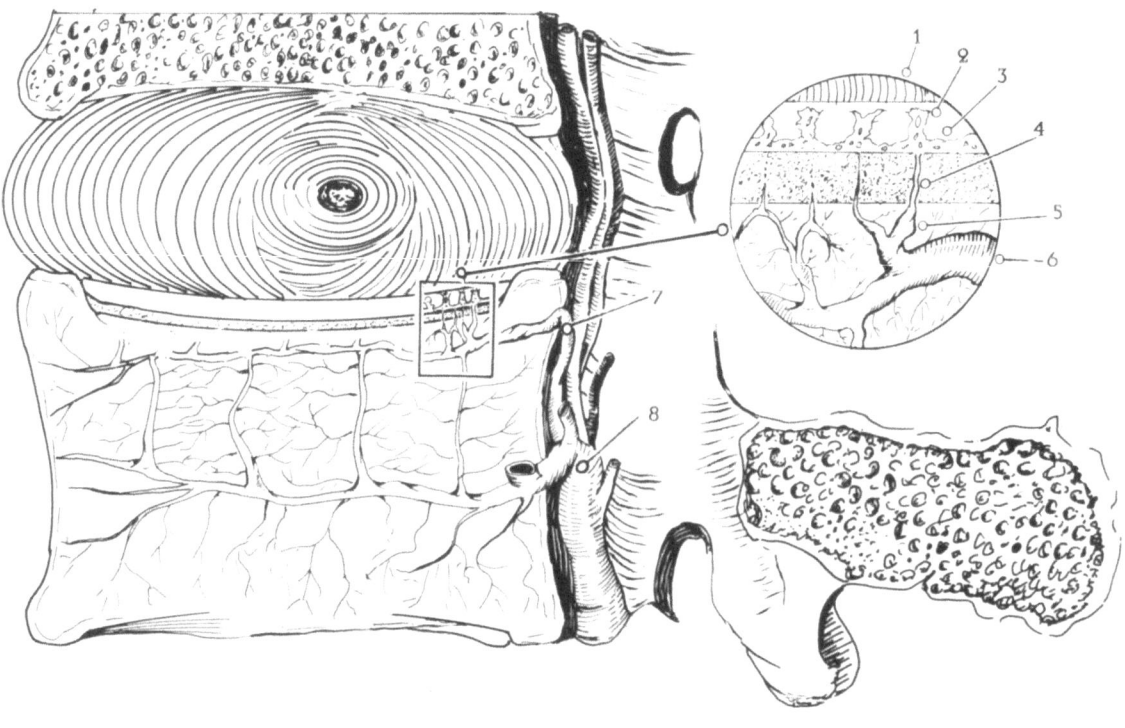

figure 6.13

A schematic drawing to show the spatial relationships of the veins of a typical vertebral body. 1, Intervertebral disc; 2, capillary bed in vertebral end-plate cartilage; 3, subchondral postcapillary venous network on the vertebral end-plate; 4, vertebral end-plate perforated by short vertical venous tributaries; 5, vertical tributary from the subchondral postcapillary venous network, draining to the horizontal subarticular collecting vein; 6, horizontal subarticular collecting vein; 7, horizontal subarticular collecting vein joining the anterior internal vertebral venous plexus; 8, basivertebral vein joining the anterior internal vertebral venous plexus; and 9, vertical tributary of the basivertebral system of veins (vertical vein of the vertebral body). (Reproduced by courtesy of the editor, *J. Bone Joint Surg.* 55-B, 1973, and J. B. Lippincott and Company from *Clinical Orthopaedics and Related Research*, No. 115, 1976.)

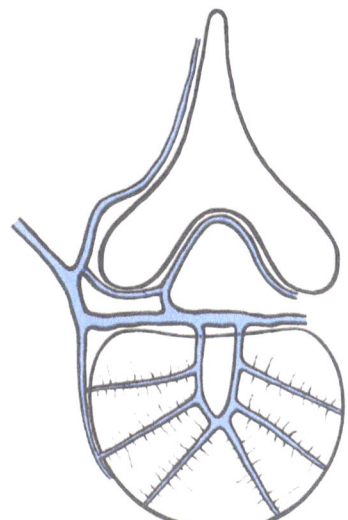

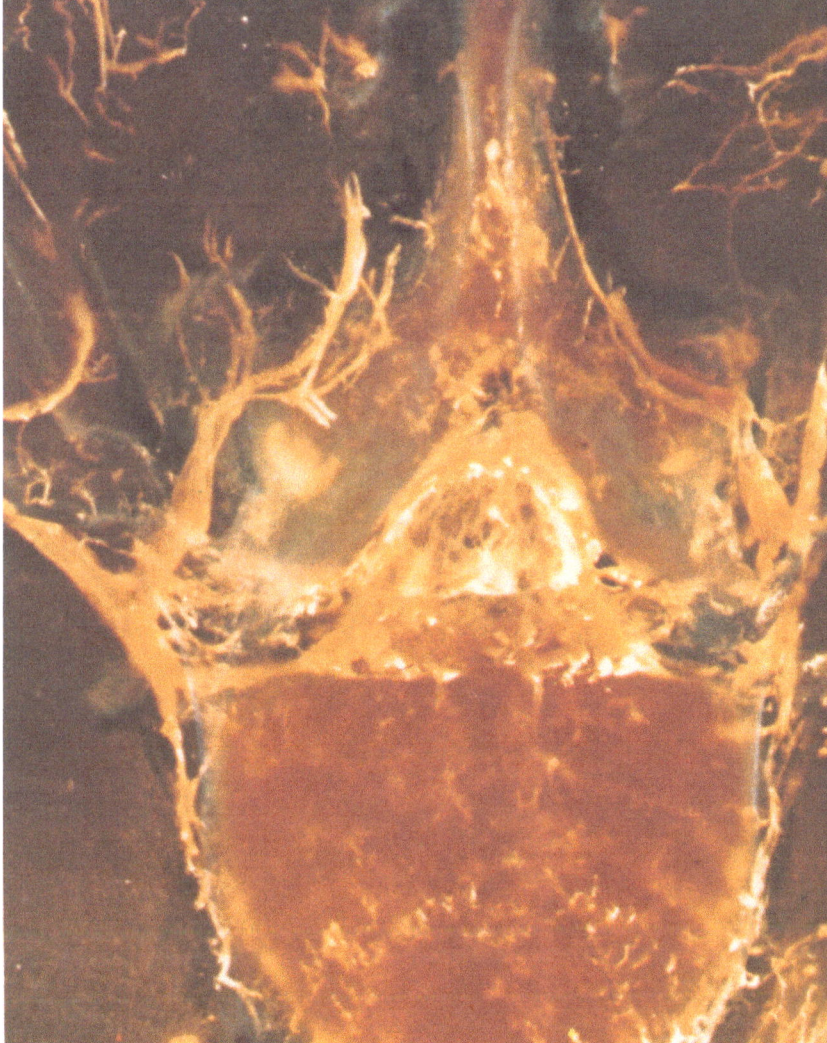

figure 6.14

A photograph of a transverse section through the middle of an upper
lumbar vertebral body and including the lamina, spinous process, and
facet joints posteriorly. Spalteholz cleared specimen in which the
veins have been distended. Photographed by transmitted light. An
explanatory schematic drawing is found alongside. The relationships of
the internal vertebral venous plexus to the basivertebral system of
Batson can be seen. In addition the anastomoses between branches
of the external vertebral venous system and the internal vertebral
venous system are shown.

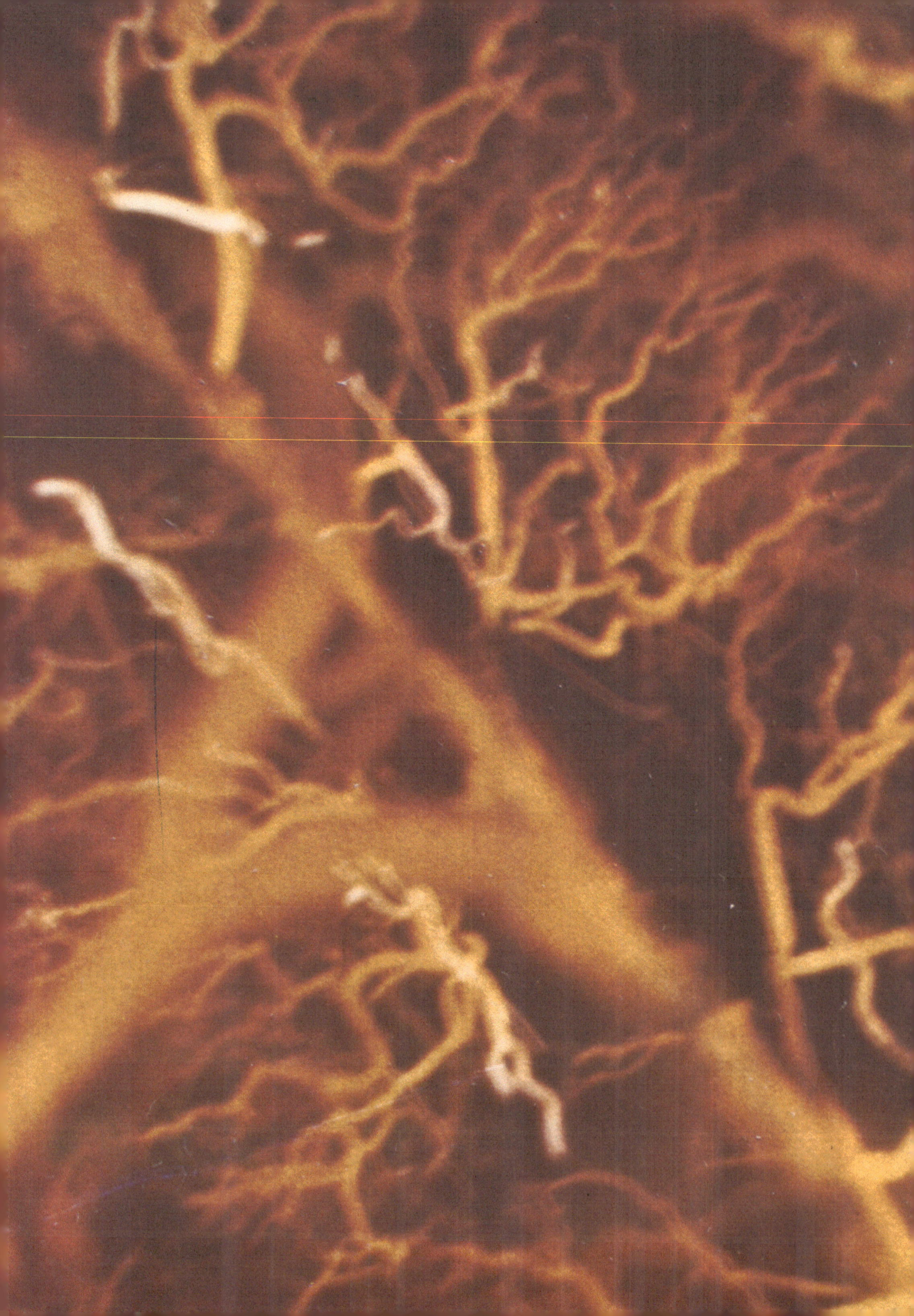

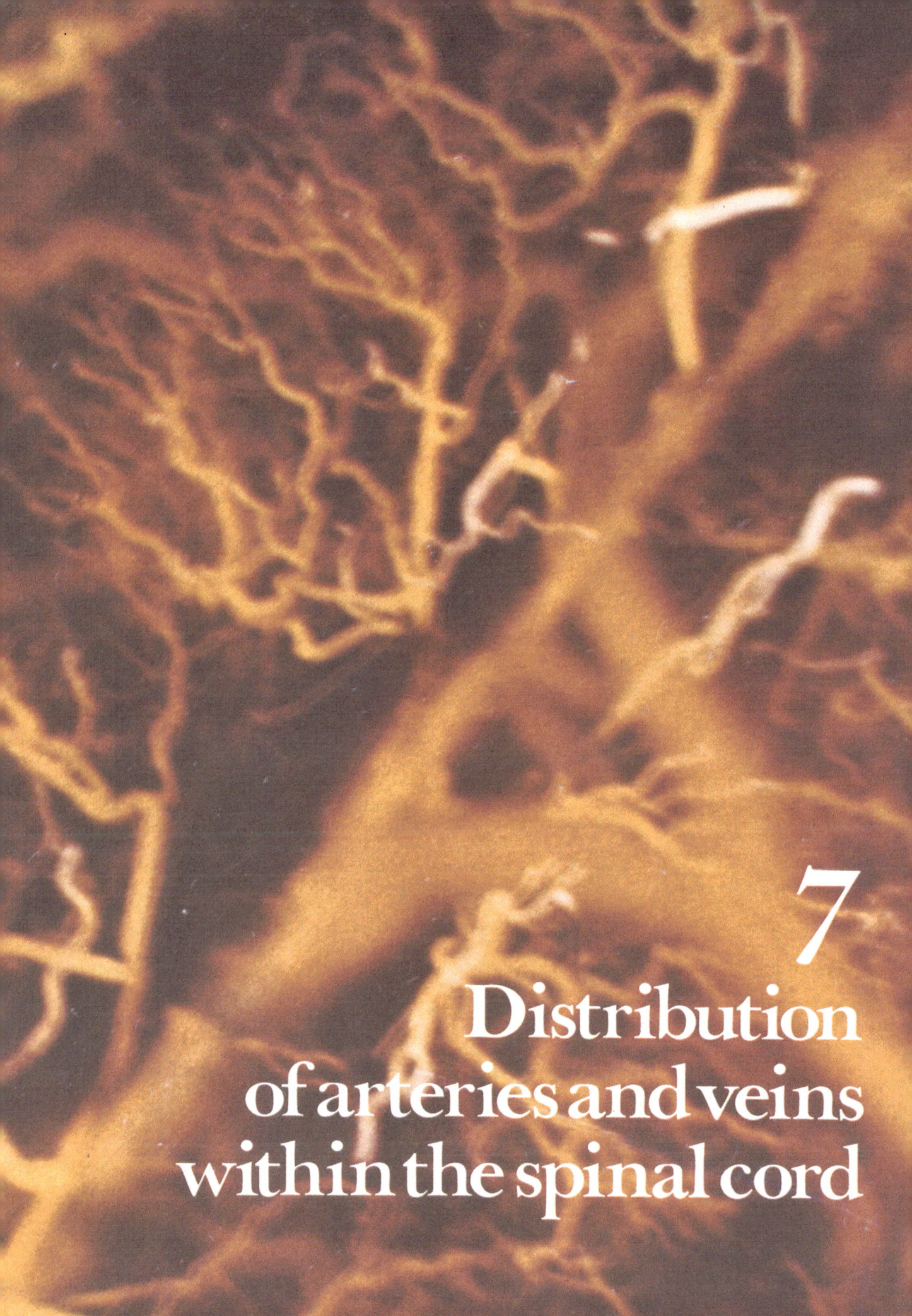

7

Distribution
of arteries and veins
within the spinal cord

Although many studies on the blood supply of the spinal cord have been published since the 19th century, there are still deficiencies in the descriptions of vessels within the cord. One of the major contemporary papers written by Turnbull, Brieg, and Hassler (1966) contains some fine illustrations of the intrinsic arteries of the cervical cord. These authors provide no illustrations of the venous drainage of this segment and they make a number of unsubstantiated comments about the density of veins within the cord. Recently Dommisse (1975) has published an important monograph, *The Arteries and Veins of the Human Spinal Cord from Birth*, yet even in this work there are no illustrations of the vessels within the spinal cord.

In this chapter we have selected a series of photographs to illustrate the salient features of the arterial supply and venous drainage of the spinal cord in man. Our methods of specimen preparation have allowed us to produce good quality injections on the arterial side to arteriolar levels, but on the venous side retrograde filling only to the level of small venules. We have no competence therefore to comment on capillary bed formation within the cord.

Close study of the illustrations presented here reveals interesting differences in the patterns of distribution of small arteries in the cervical, thoracic, and lumbar areas. The claim that the blood supply to the thoracic cord is tenuous cannot be supported.

THE ARTERIES OF THE SPINAL CORD

The spinal cord is penetrated around its circumference posteriorly by radially disposed arteries, branches of the *posterolateral longitudinal arterial trunks* and of the pial plexus. Anteriorly, based on the *anterior median longitudinal arterial trunk*, the cord is penetrated by a series of horizontally orientated central arteries. The number of these varies, being greatest in the cervical and lumbar enlargements. In the cervical area they pass backward along a slightly oblique path, alternate branches usually passing to left and right just anterior to the central canal of the cord. Each branch then bifurcates to spread out laterally into the anterior column of gray matter in the pattern of branches of a closely cropped leafless tree.

In the region of the lumbar enlargement, the central arteries are most numerous. They pass backward in the horizontal plane in slightly wavy courses, being somewhat longer than their counterparts in the cervical region.

In the thoracic region the central arteries are more widely spaced at their origins along the anterior median longitudinal arterial trunk of the cord. They course backward obliquely and break up into branches which form long ascending and descending loops within the anterior horns of the cord, like grapevines after their leaves have fallen. In sagittal section, the main stems of the thoracic central arteries form wedge-like patterns with arteries penetrating the posterior surface of the cord. This gross pattern is seen also in the venous drainage. By contrast, in sagittal sections in the cervical and lumbar segments the anterior and posterior penetrating arteries lie in a straight line in the horizontal plane.

THE VEINS OF THE SPINAL CORD

In transverse sections, the spinal cord is drained by radiate veins which accompany the entrant arteries. There are a number of major differences between the two systems of vessels. Large anterior and posterior median veins are found, the latter having no real arterial counterpart. We have found that there is often a continuous venous channel formed in the midline by confluence of these anterior and posterior median veins, which deviates to one side around the central canal of the cord. Kadyi (1889) and Herren and Alexander (1939) have described centrodorsolateral venous anastomoses linking the anterior median spinal veins with veins on the posterolateral surface of the cord.

The median veins, if separate, each bifurcate close to the central canal, collecting tributaries from both sides of the gray matter. In turn, the veins emerge onto the surface of the cord to join, respectively, the anterior median longitudinal venous trunk and the posterior median longitudinal venous trunk. The other radiate veins join the venous pial plexus around the circumference of the cord.

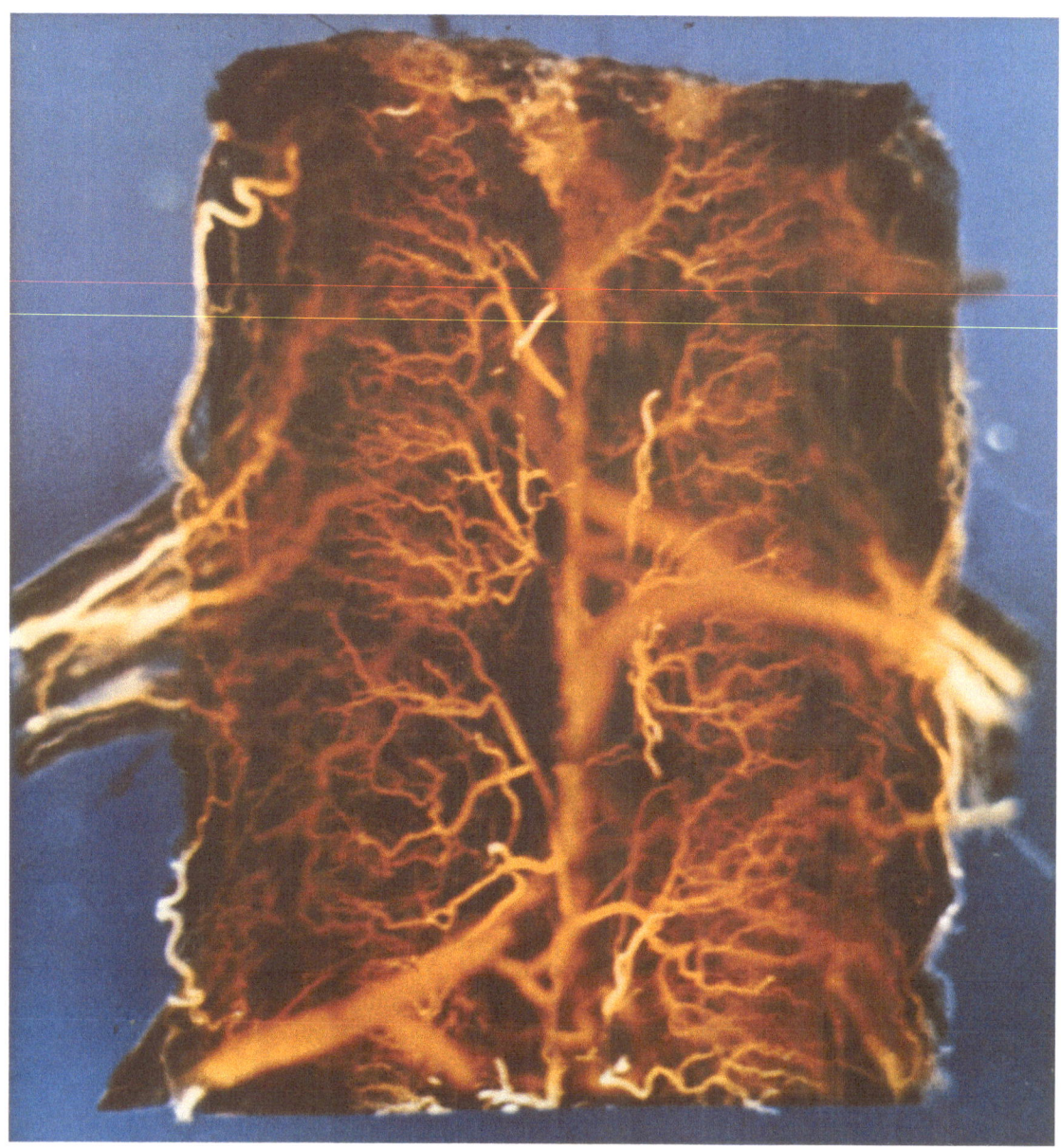

figure 7.1

A photograph of a coronal section of a segment of the cervical spinal
cord from a child aged 13 years. The section is cut just posterior to the
bulbous anterior horns of gray matter. In the midline *the anterior
median longitudinal arterial trunk* of the spinal cord can be seen out of
focus. The specimen has been prepared to illustrate the manner of
branching of *the central arteries* of the spinal cord. Note the deviation
of each central artery to one or other side of the cord where it
branches laterally in the pattern of a closely cropped leafless tree.
Occasionally a single central artery will bifurcate to form a treelike
pattern on the right and left side of the midline.

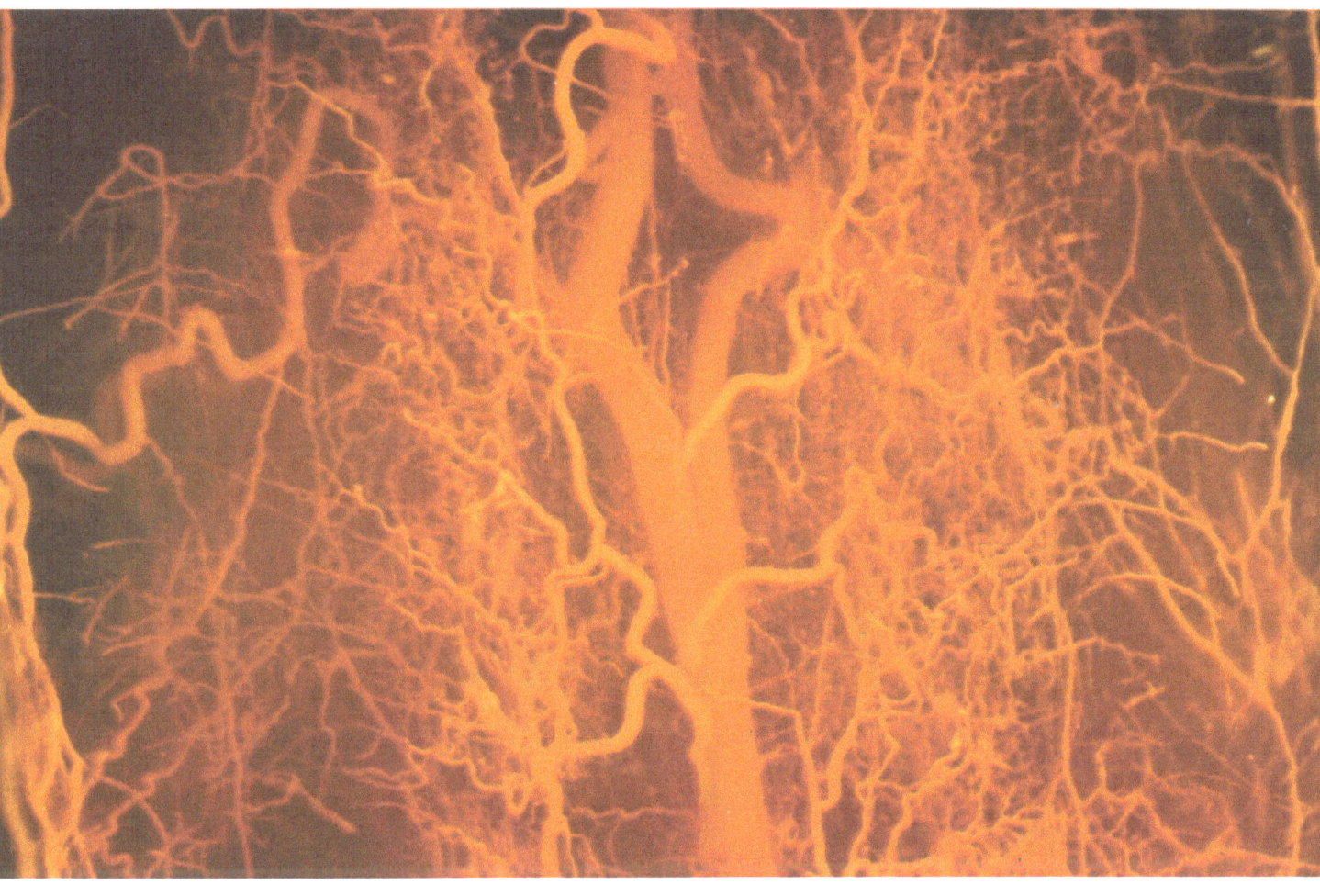

figure 7.2

A photograph of a coronal section of a segment of the cervical spinal
cord from a female aged approximately 35 years. Arterial injection,
Spalteholz cleared specimen. The origins of four central arteries of the
spinal cord from the *anterior median longitudinal arterial trunk of the
cord* can be seen in the midline from above downward. The first
deviating toward the left side of the photograph before breaking up
into treelike branches, the second and third arteries deviating to the
right side and the fourth again deviating to the left side.

On the extreme left side of the photograph a large pial plexus artery
can be seen.

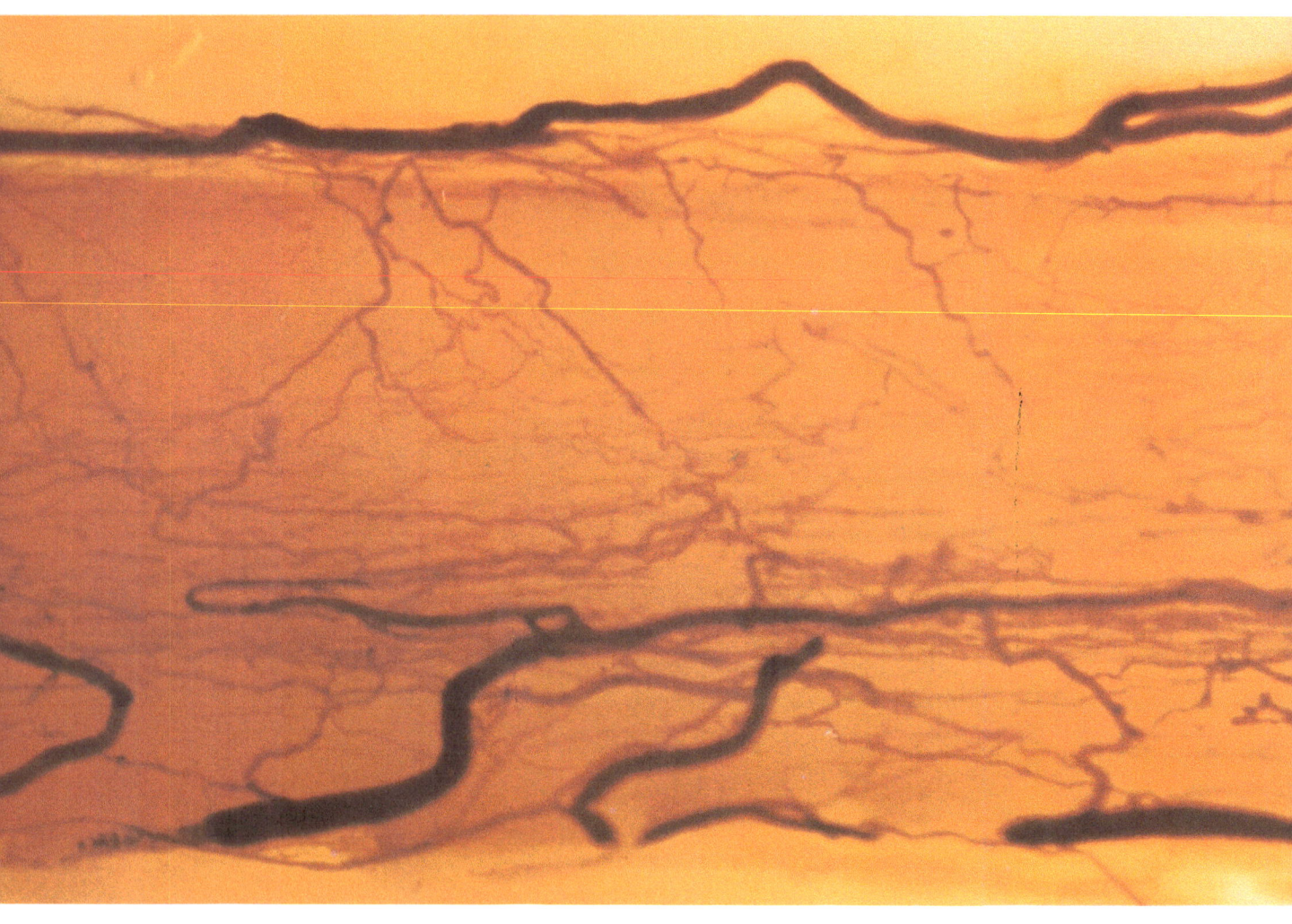

figure 7.3

A photograph of a thin median sagittal section from the thoracic spinal
cord of a female aged 35 years. Arterial injection, Spalteholz cleared
specimen. Note the obliquity of the central arteries in the lower part of
the picture. The branching system differs markedly from that seen
in the cervical region in Figures 7.1 and 7.2. Note also the wedgelike
patterns formed by the penetrating arteries from the posterior surface
of the cord where they meet the anterior central arteries in the region
of the central canal of the spinal cord.

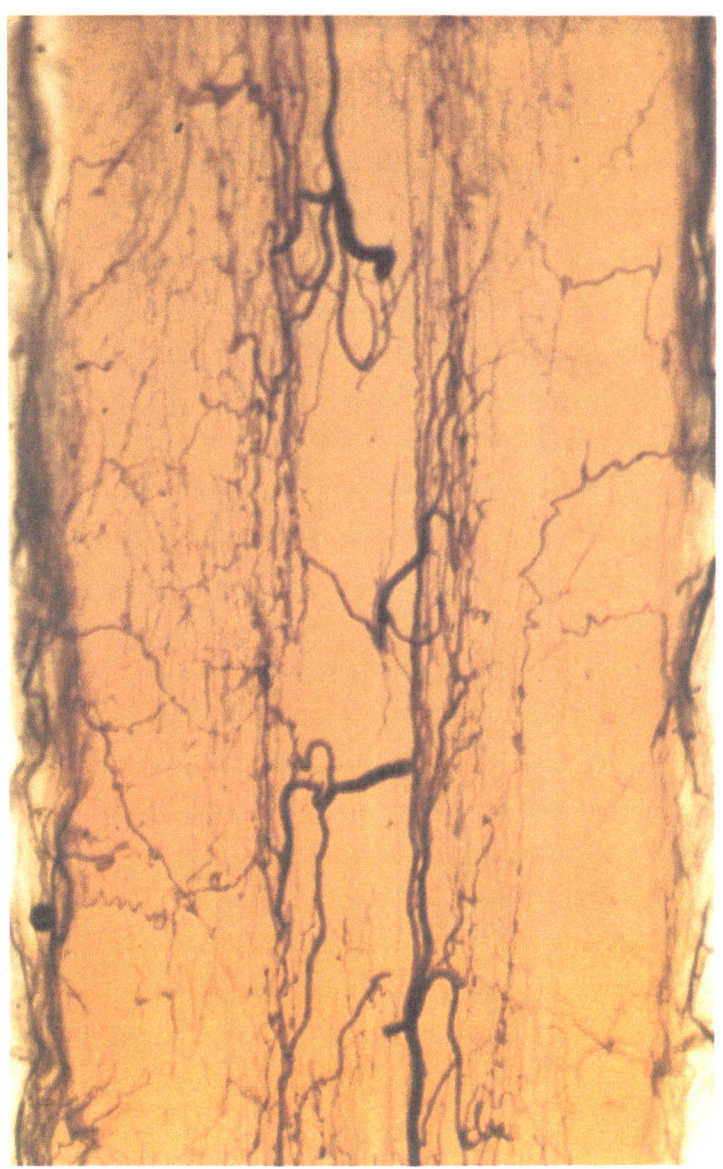

figure 7.4

A photograph of a thin coronal section from the thoracic spinal cord of
a female aged 35 years. Arterial injection, Spalteholz cleared specimen.
The pattern of branching of four *central arteries* can be seen, alternate
vessels passing to the right and left of the midline from above
downward. The branching system of these vessels differs from that
seen in the other areas of the cord. Note the long ascending and
descending loops which are formed in the pattern of grapevines after
their leaves have fallen.

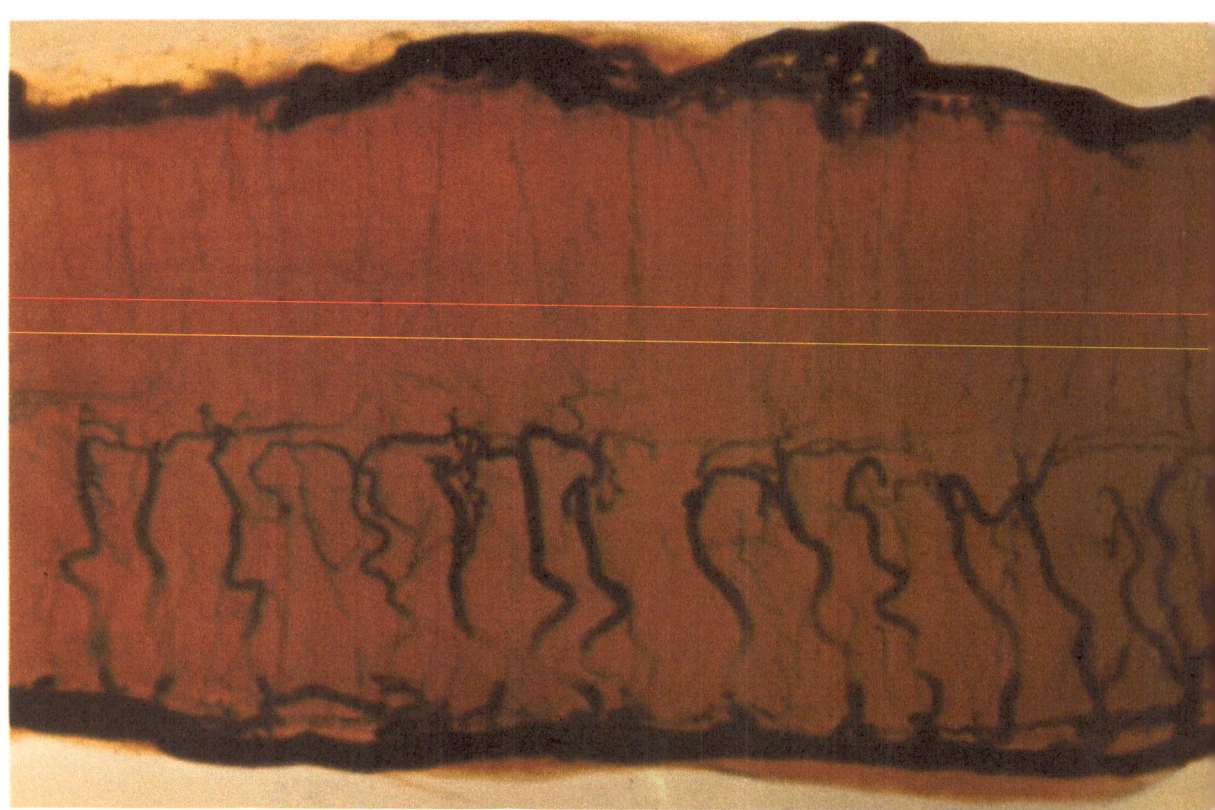

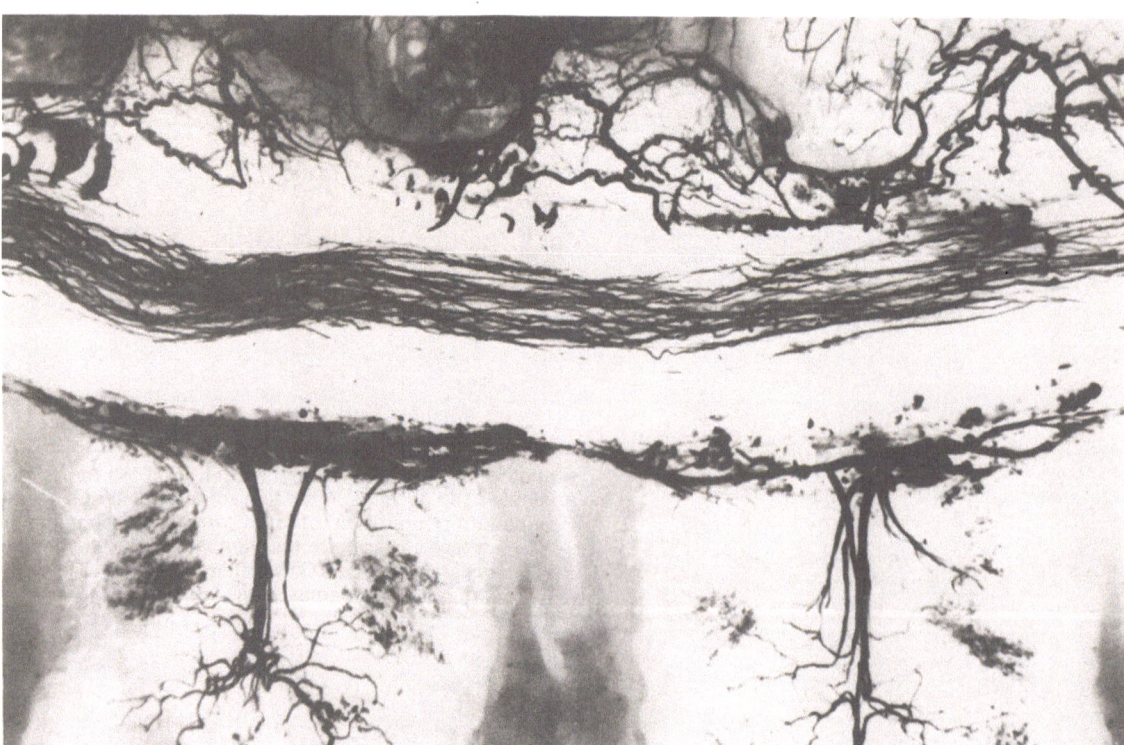

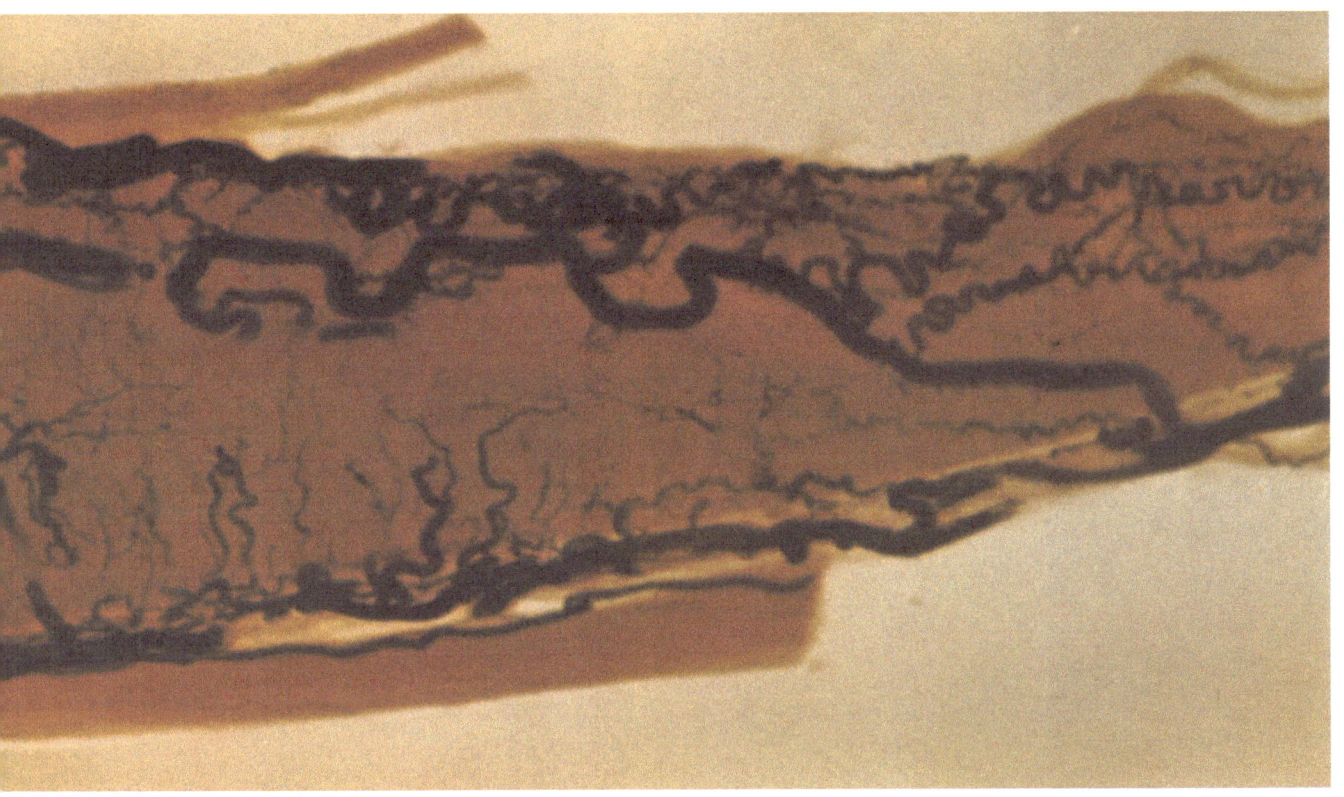

figure 7.5

A photograph of a thin median sagittal section of the lumbar
enlargement and conus medullaris from an adult spinal cord. Arterial
injection, Spalteholz cleared specimen. The anastomosis between the
posterolateral longitudinal arterial trunk of the spinal cord with the
anterior median longitudinal arterial trunk can be seen on the right
extremity of the photograph. Note the density of the *central arteries* of
the cord which run backward along slightly wavy courses in the
horizontal plane. Arteries penetrating from the posterior surface of the
cord can be seen running forward in the horizontal plane. These are of
very narrow caliber.

figure 7.6

A radiograph of the upper lumbar spine from an adult shows the
density of small arteries in the cauda equina. These vessels run a long
course. Their sizes range from about 30 to 70 μm. The cauda equina
vessels are well filled, but note that the intraosseous arteries of the
two vertebral bodies are only partly filled. The large centrum arteries
in both of these vertebrae are well filled.

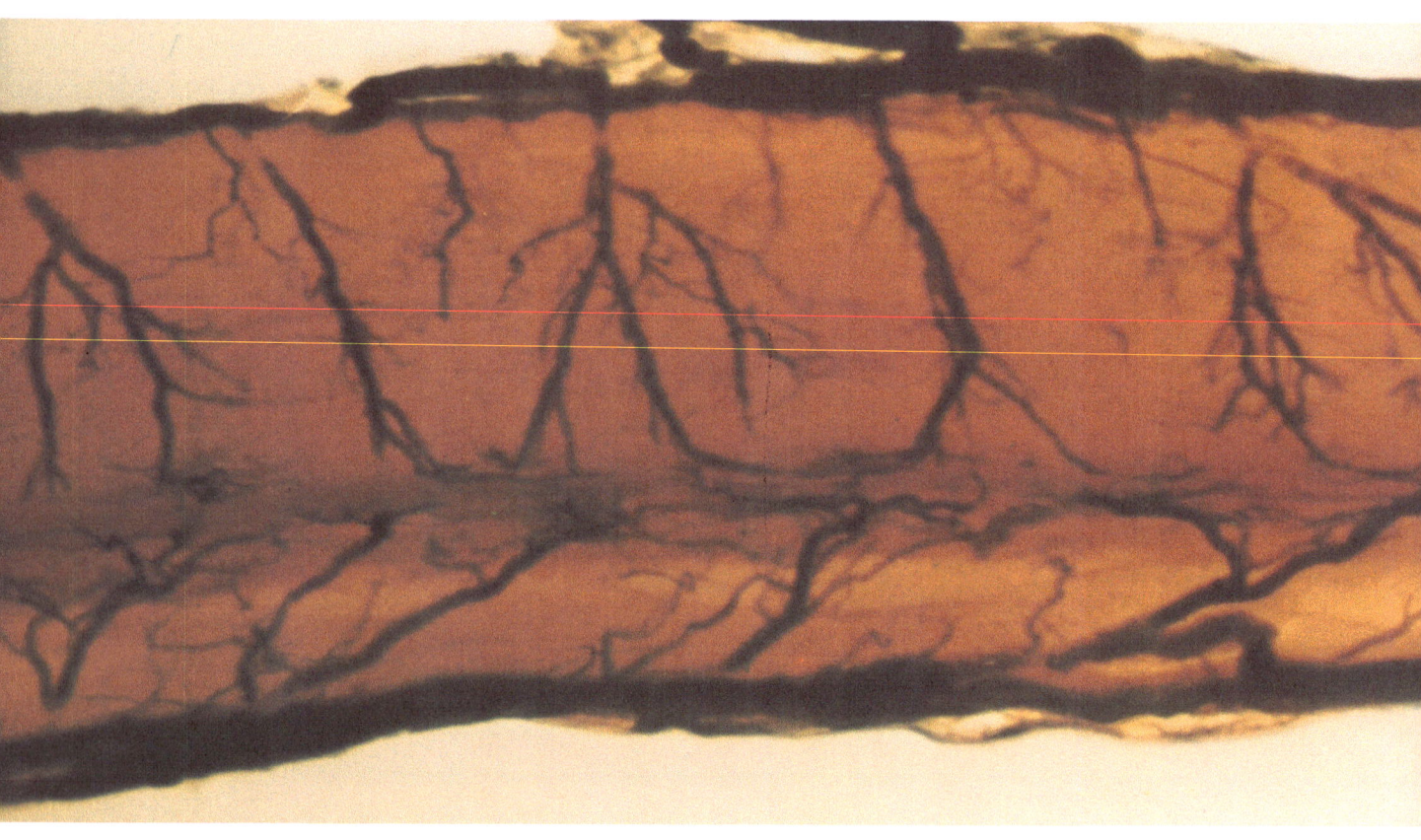

figure 7.7

A photograph of a thin median sagittal section from the thoracic spinal
cord of a male aged 21 years. Venous injection, Spalteholz cleared
specimen. The veins of the spinal cord have a characteristic pattern
being of larger caliber than the corresponding artery, with venous
radicles entering the main stems obliquely.

Note the wedge-shaped pattern formed between the main stems of the
central veins and the main stem of the veins in the posterior segment
of the cord. Compare this with Figure 7.3, showing the similar spatial
distribution of the arteries in a thoracic spinal cord.

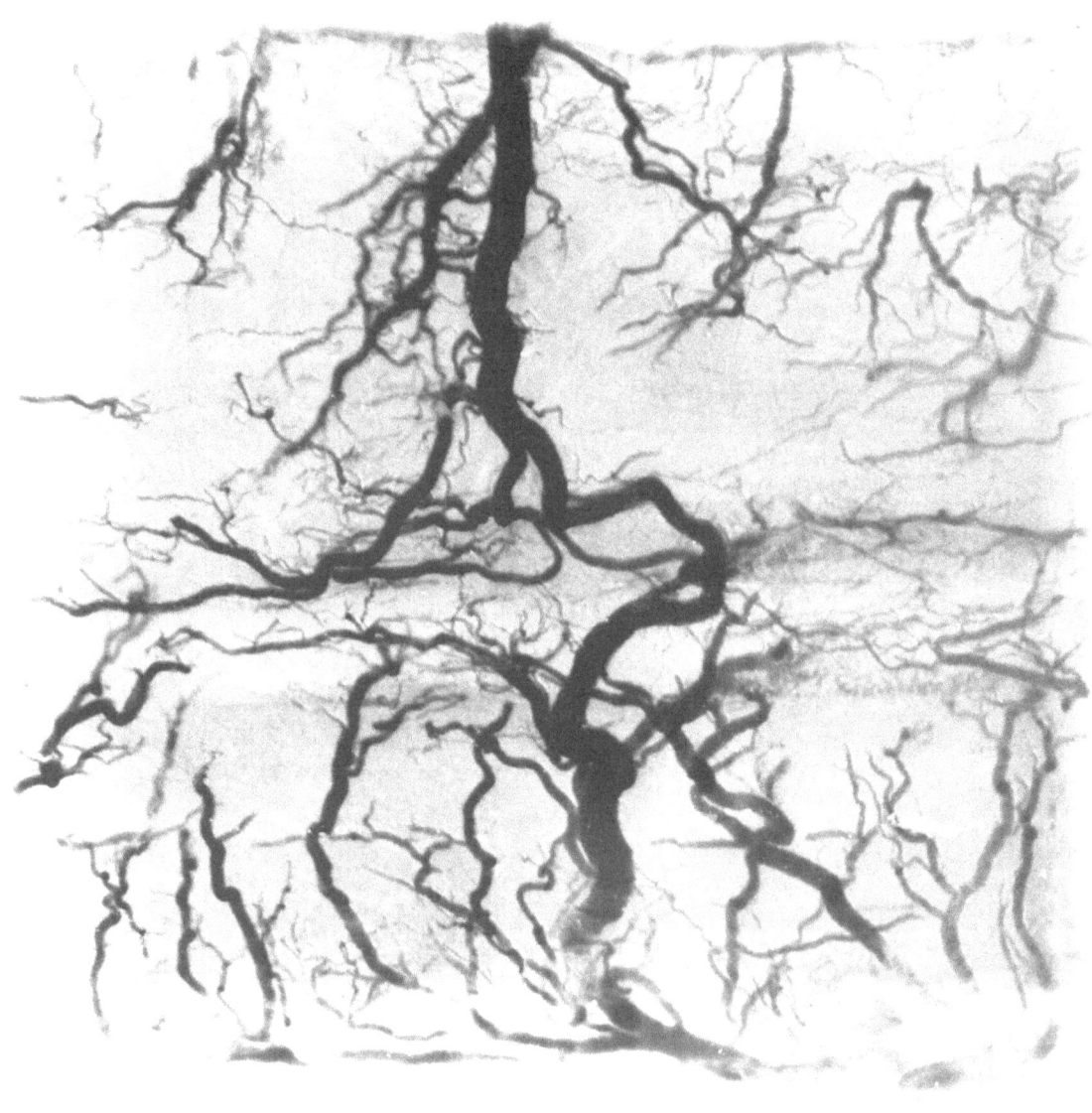

figure 7.8

A photograph of a median sagittal section from a segment of the
thoracic spinal cord of a male aged 13 years. Venous injection,
Spalteholz cleared specimen.

At the bottom of the picture in the center, a large tortuous *central
vein* passes backward deviating in the region of the central canal of
the cord to pass backward in continuity with the *posterior median vein*
of the spinal cord. This important venous channel does not appear to
have been described by previous authors. Other examples of it are seen
in Figures 7.9, 7.10, and 7.11a and b.

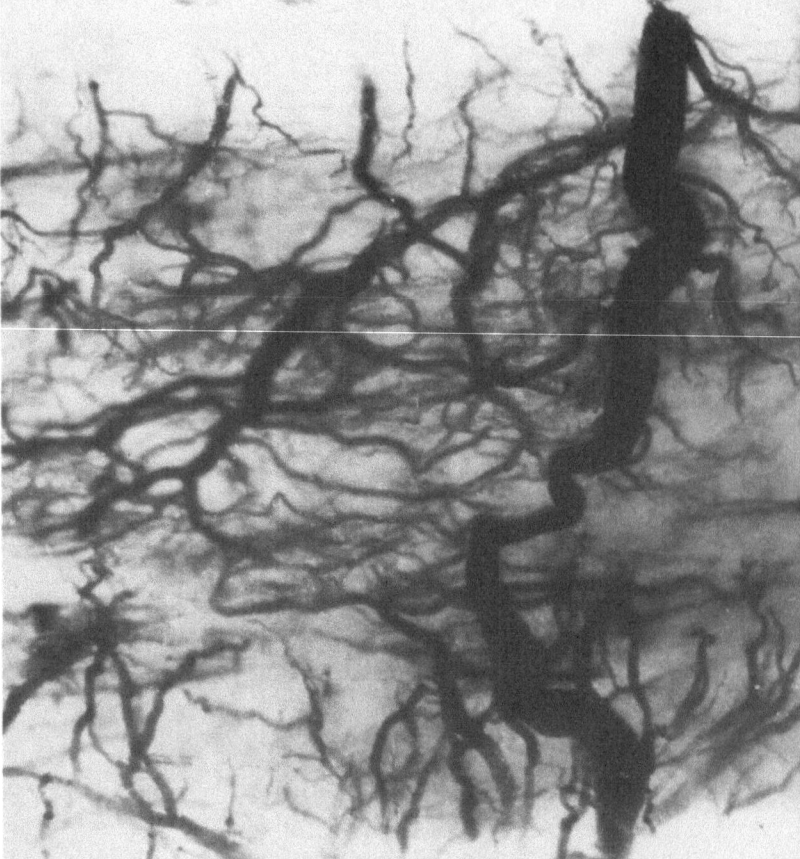

figure 7.9

A photograph of a median sagittal
section from another part of the
spinal cord from the specimen
illustrated in Figure 7.8. A large
central venous anastomosis is
shown.

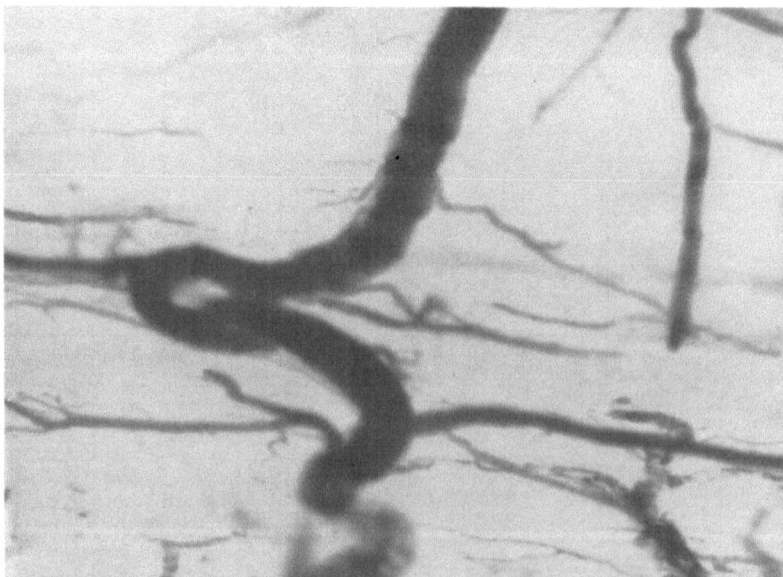

figure 7.10

A detailed photograph of a *median
anteroposterior venous anastomotic
channel* from the spinal cord of a
female aged 67 years cut in sagittal
section. Spalteholz cleared specimen.

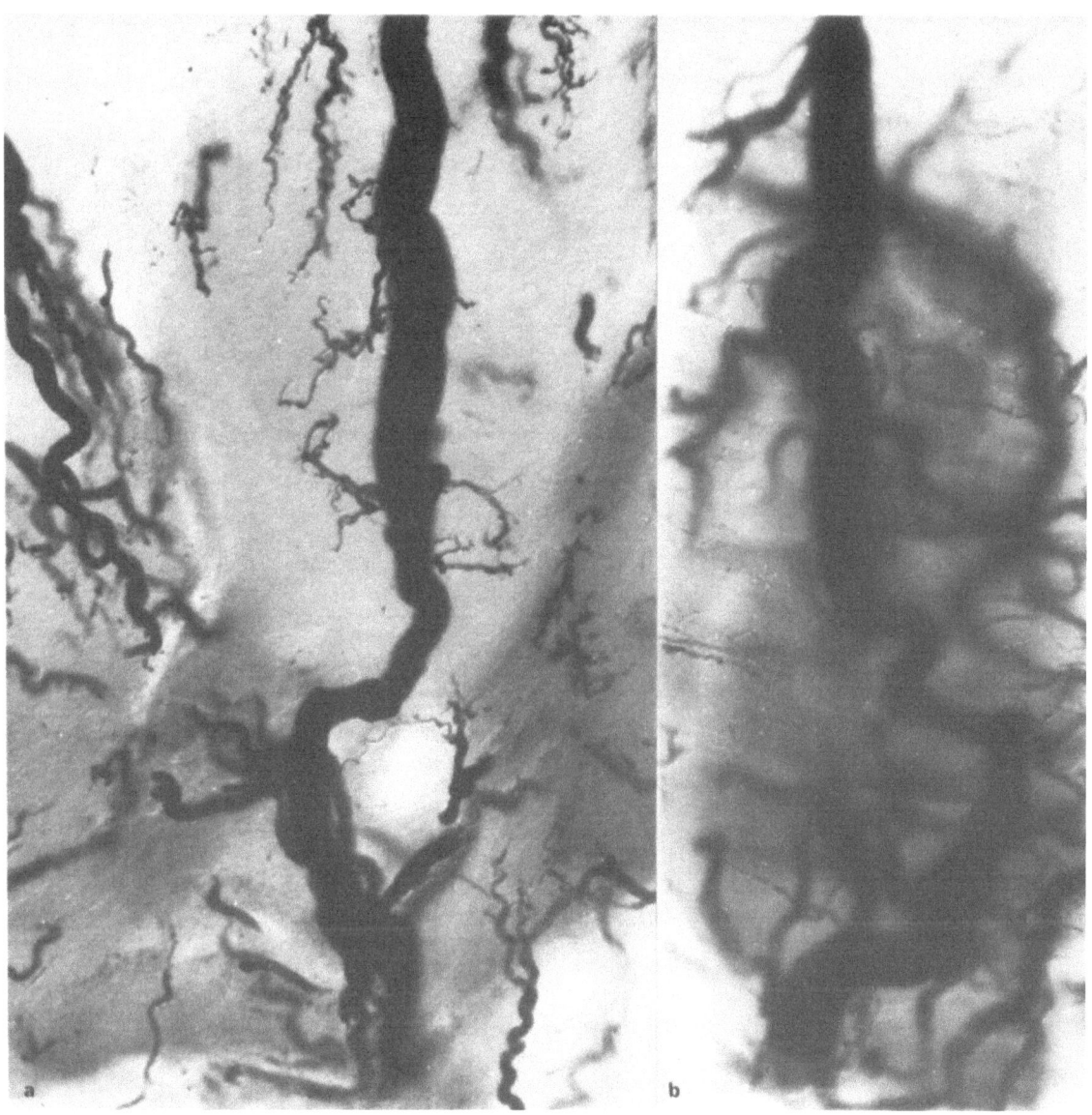

figure 7.11a and b

(a)
A transverse section through the midthoracic region of the spinal cord
illustrated in Figures 7.8 and 7.9. The relationship of the *median
anteroposterior venous channel* of the spinal cord to the central canal
of the cord can be seen.

(b)
Alongside is a lateral view of the same vein.

The blood supply of the vertebral column
and spinal cord

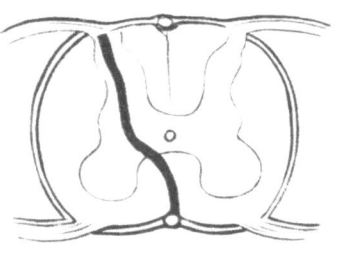 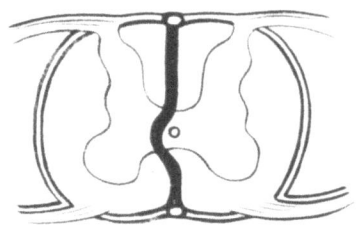 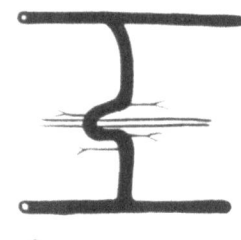

a   b   c

figure 7.12a,b, and c

Line drawings to illustrate the major venous anastomoses within the
spinal cord.

(a)
The *centrodorsolateral venous anastomosis* described by Kadyi,
Herren, and Alexander.

(b) and (c)
The *median anteroposterior venous channel* of the spinal cord viewed
in transverse and sagittal section as described in the present work by
Crock and Yoshizawa.

figure 7.13

A photograph of a median sagittal section from the spinal cord
of a male aged 13 years. Venous injection, Spalteholz cleared
specimen. The section is cut at the level of the dorsolumbar junction.
Note the density of the veins, their gross spatial arrangement resembling
that of the entrant arteries.

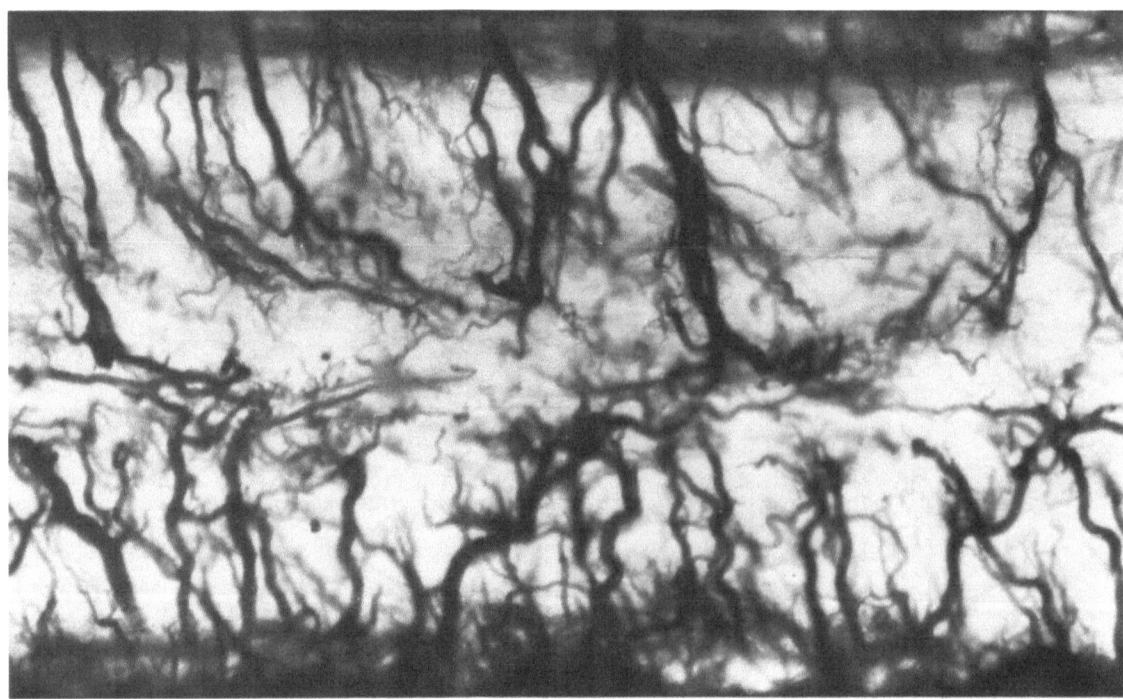

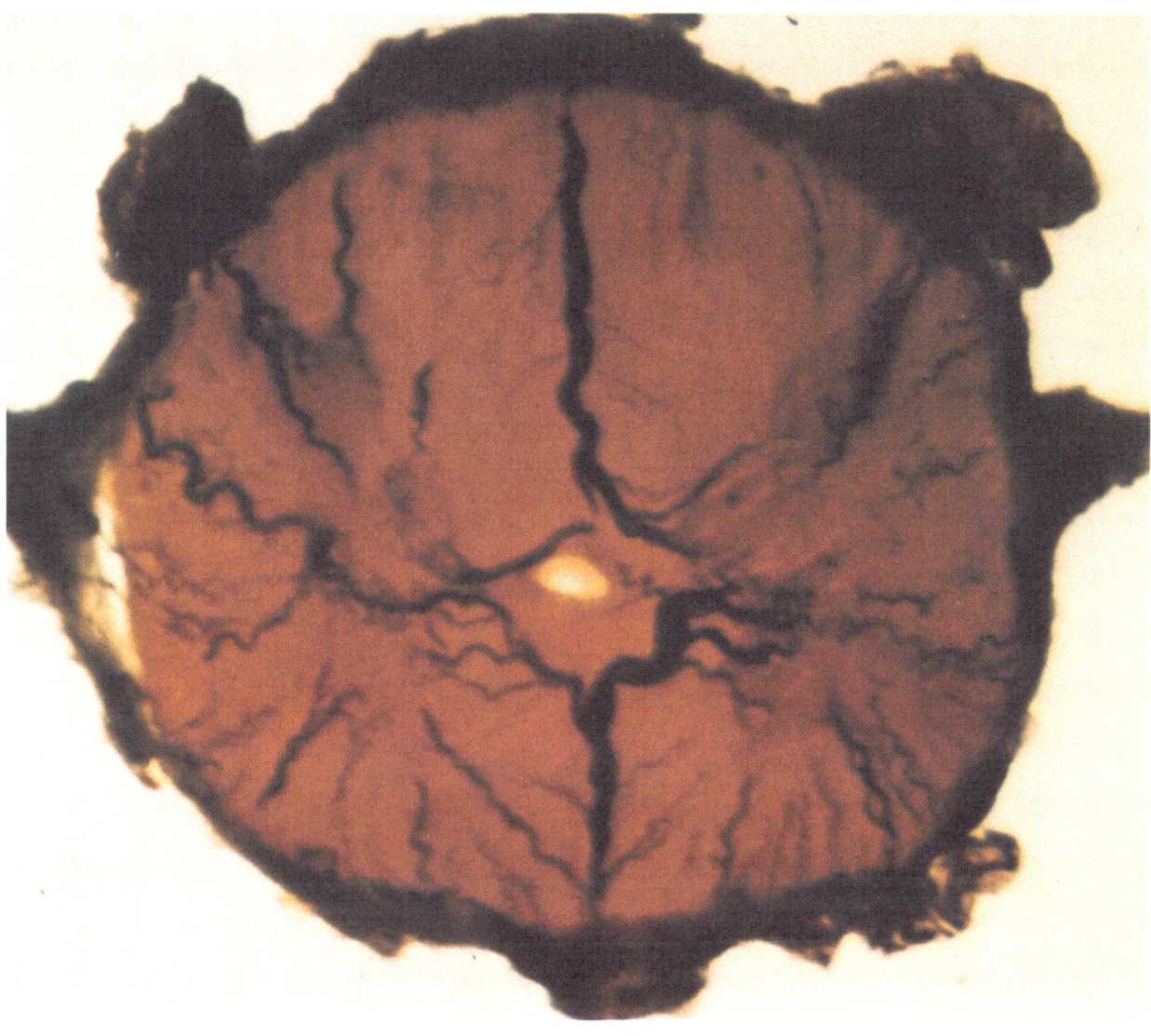

figure 7.14

A photograph of a thin transverse section from the lumbar cord of a
child aged 13 years. Spalteholz cleared specimen, venous injection.
The *anterior and posterior median veins of the cord* are seen
bifurcating in close relation to the central canal of the cord, draining
the central gray matter on each side. Elsewhere in the specimen the
radiate veins of the cord can be seen draining to the pial plexus.

figure 7.15

A photograph of a thin transverse
section through the thoracic spinal
cord from the specimen illustrated
in Figure 7.14. The overall radiate
pattern of venous drainage can be
seen. Note the large *posterior median
vein* draining the posterior horns of
gray matter on both sides.

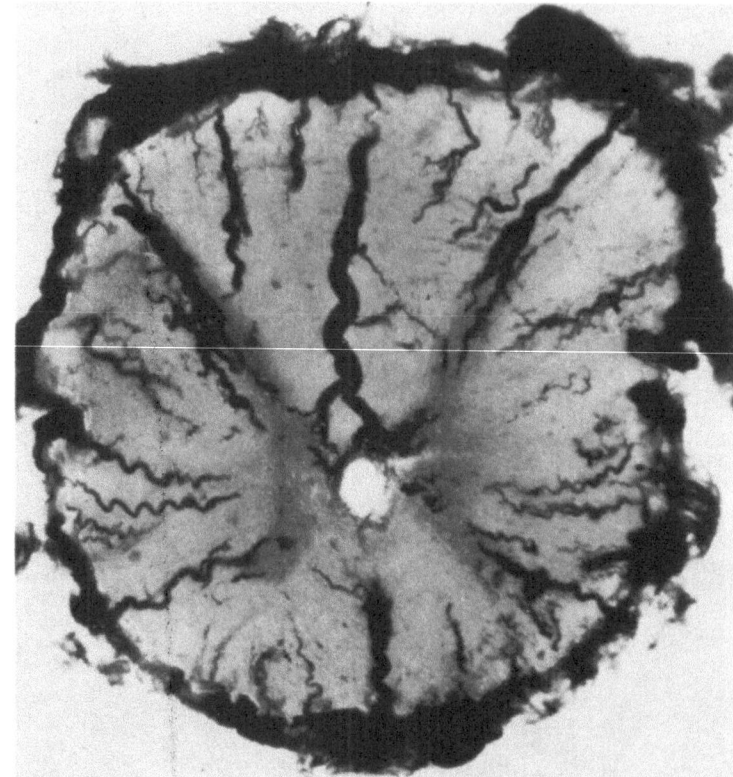

figure 7.16

A photograph of a thin transverse
section through the lumbar cord of
the same specimen illustrated in
Figures 7.14 and 7.15. Gross details
of the anterior central vein of the
cord are seen in this specimen with
branches draining the medial aspects
of both anterior gray columns.

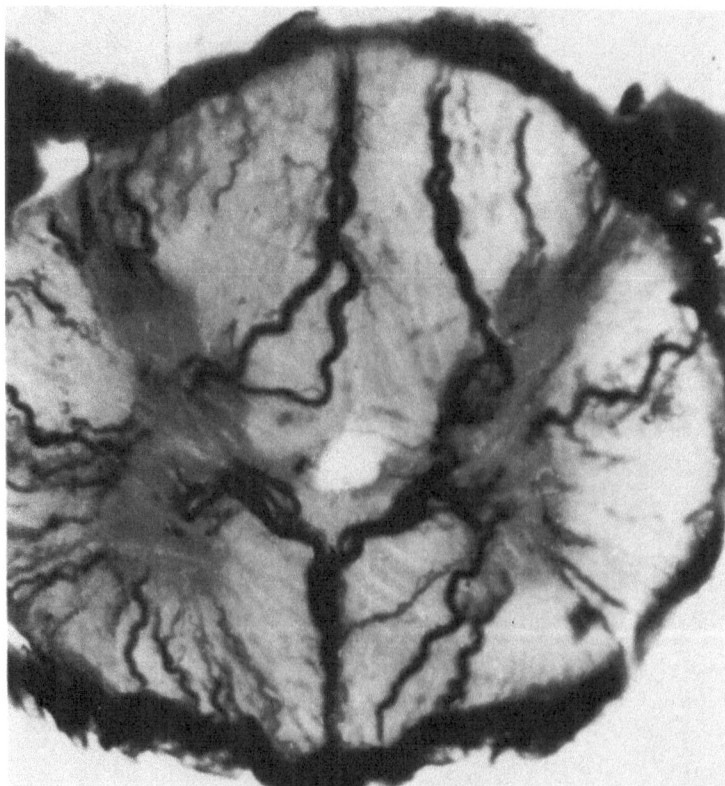

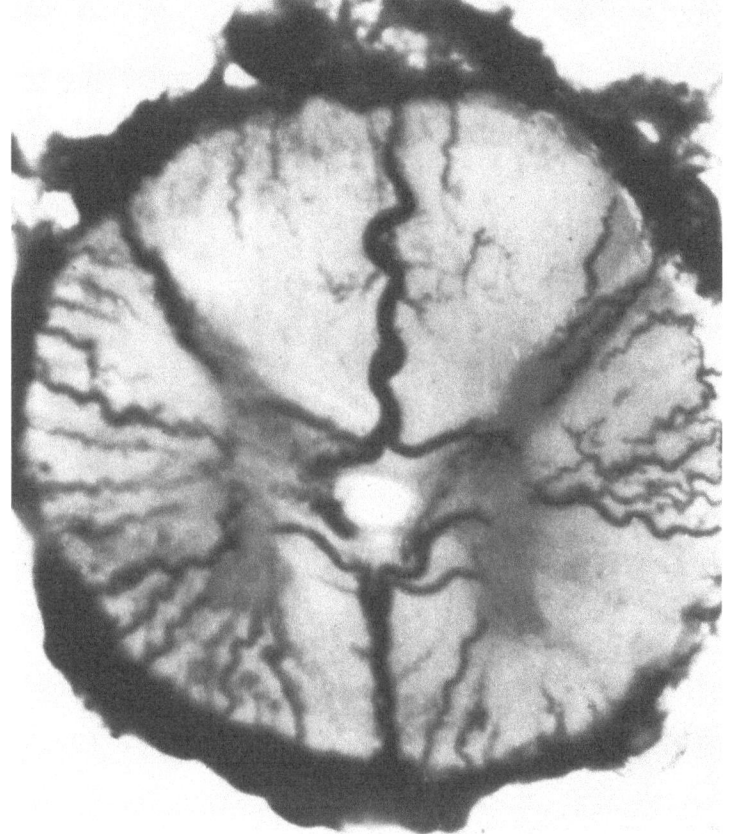

figure 7.17

A photograph of a thin transverse
section through the lower thoracic
cord of the specimen illustrated in
Figures 7.14, 7.15, and 7.16, showing
the *anterior and posterior central
veins* of the spinal cord which
commonly bifurcate close to the
central canal to drain, respectively,
the anterior and posterior columns of
gray matter. *Radiate veins* are shown
well filled on the left side of the
photograph.

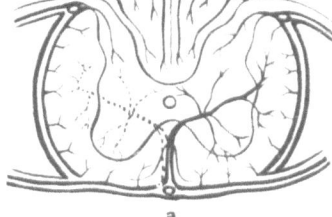

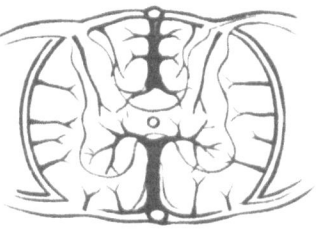

figure 7.18a and b

Composite diagrams to illustrate the basic differences in gross spatial
distribution of arteries and veins within the spinal cord.

(a)
In the solid outline an artery deviating to the right of the diagram
is shown, and on the left in dotted outline, another central artery is
indicated deviating to the left side of the anterior gray column.

(b)
depicts the large *anterior and posterior median veins* of the cord, each
bifurcating close to the central canal to drain the inner aspects of the
gray columns. Note the differing dimensions of the veins when
compared with the arterial patterns in (a).

NOTES ON MATERIALS AND METHODS

The studies presented in this work, *The Blood Supply of the Vertebral Column and Spinal Cord in Man*, are based on specimens prepared from injections made at necropsy, usually within 12 hours of death. Without exception, injections have been made either through a major artery such as the aorta or a major vein, usually the azygos. We believe that inferences drawn about vascular patterns in the spinal cord injected after removal of the cord segment may be misleading.

Vessels have been filled in most instances with Micropaque (Damancy and Company, Hertfordshire, England and Nicholas Pty. Ltd., Melbourne, Australia) and in some cases colored epoxy resin or colored latex rubber solution.

For the venous injections, a metal cannula of 2 mm diameter was tied into the azygos vein just proximal to its point of entry into the superior vena cava. The superior and inferior vena cava were ligated at heart level. Opening the abdomen was avoided if possible. However, if the abdominal viscera were to be removed before injection, the inferior vena cava was ligated just below the liver. The renal veins were separately ligated. After removal of the rectum, the bladder was opened and filled with liquid nitrogen, as was the pelvis. If there was any vertebral injury above the level of the azygos vein, the vertebral column was divided and one vertebral body removed. Into this defect liquid nitrogen was poured until the whole area was frozen. In effect, we have used liquid nitrogen to snap-freeze the vertebral column in segments, thus establishing a state equivalent to that in the lower limb after the application of a tourniquet before beginning venous injections. This simple technical device has allowed us to obtain venous injections of good quality both in the spinal cord and the bones of the vertebral column (Crock, Yoshizawa, and Kame, 1973).

The volume of Micropaque used in an adult of average size for this study has varied between 3 and 5 liters delivered for 30 to 60 minutes at 5 to 10 lbf/square inch (35 to 70 kgf/square centimeter or N/square meter), using dry nitrogen as the propellant gas.

After the injection of Micropaque, the vertebral column was removed complete with attached muscles. Specimens were placed in a deep-freezing refrigerator at −40°C for approximately 36 hours. Sections of the vertebral columns were then cut in sagittal coronal and horizontal planes with an electrically powered industrial meat-cutting band saw. After fixation in 10% buffered formalin, the specimens were decalcified and cleared by a modified Spalteholz method, excluding the use of benzol.

Finally selected, the specimens were x-rayed with a Softex type E.S.M., super soft x-ray apparatus using Agfa-Gevaert graphic film R.O.81, p. Some were photographed by transmitted or reflected light in clearing fluid (Spalteholz, 1911), with the use of Kodak Ektachrome film ASA 125, or some of the Kodak fine grain black and white films.

Our material includes specimens obtained from 125 cadavers at postmortem with ages ranging from newborn to 80 years. Seventy-three spinal cords have been used in the study, 33 for arterial specimens and 44 venous injections. In the studies on spinal cord circulation it is important to note that the whole vertebral column has been fixed in formalin in every case and processed through Spalteholz's method, before any cord dissection has been attempted. In this way we have been able to prepare specimens without risk of damaging certain fine segmental vessels.

BIBLIOGRAPHY

Adamkiewicz, A. A. (1881a): Ueber die mikroskopischen Gefässe des menschlichen Rückenmarkes. Trans. Int. Med. Cong., 7th session, London, 1: 155–157.

Adamkiewicz, A. A. (1881b): Die Blutgefässe des menschlichen Rückenmarkes. 1. Die Gefässe der Rückenmarkssubstanz. *Sitzungsber. Akad. Wiss. Wien Math.-Naturwiss. Kl.* 84: 469–502.

Batson, O. V. (1940): The function of the vertebral veins and their role in the spread of metastases. *Ann. Surg.* 112: 138–149.

Crock, H. V. (1967): *The Blood Supply of the Lower Limb Bones in Man.* Edinburgh and London: E. & S. Livingstone Ltd.

Crock, H. V., Yoshizawa, H., and Kame, S. K., (1973): Observations on the venous drainage of the human vertebral body. *J. Bone Joint Surg.* 55-B: 528–533.

Dommisse, G. F. (1975): *The Arteries and Veins of the Human Spinal Cord from Birth.* Edinburgh and London: Churchill Livingstone Ltd.

Gillilan, L. A. (1958): The arterial blood supply of the human spinal cord. *J. Comp. Neurol.* 110: 75–103.

Gray, H. (1973): *Gray's Anatomy.* 35th Edition, Longman, Edinburgh, pp. 702–704.

Herlihy, W. F. (1947): Revision of the venous system: The role of the vertebral veins. *Med. J. Aust.* 1: 661–672.

Herren, R. Y., and Alexander, L. (1939): Sulcal and intrinsic blood vessels of human spinal cord. *Arch. Neurol. Psychiat.* 41: 678–687.

Kadyi, H. (1889): Über die Blutgefässe des menschlichen Rückenmarkes. Nach einer im XV. Bande der Denkschriften d. math. -naturw. Cl. d. Akad. d. Wissensch. in Krakau erschienenen Monographie, aus dem Polnischen übersetzt vom Verfasser. Gubrynowicz and Schmidt, Lemberg.

Spalteholz, K. W. (1911): Ueber das Durchsichtigmachen von menschlichen und tierischen Preparation; nebst Anhang: Ueber Knochenfarbung. Leipzig: S. Hirzel.

Turnbull, I. M., Brieg, A., and Hassler, O., (1966): Blood supply of cervical spinal cord in man: A microangiographic cadaver study. *J. Neurosurg.* 24: 951–965.

RECOMMENDED FURTHER READING

Adamkiewicz, A. A. (1882): Die Blutgefasse der menschlichen Rückenmarksoberfläche. *Sitzungsber. Heidelb. Akad. Wiss.* 85: 101-130.

Bolton, B. (1939): The blood supply of the human spinal cord. *J. Neurol. Psychiat.* 2: 137.

Ferguson, W. R. (1950): Some observations on the circulation in foetal and infant spines. *J. Bone Joint Surg.* 32-A: 640.

Gillilan, L. A. (1970): Veins of the spinal cord. Anatomic details; suggested clinical applications. *Neurology* 20: 860.

Guide, G., Cigala, F., and Riccio, V. (1969): The vascularization of the vertebral body in the human foetus at term. *Clin. Orthop. Relat. Res.* 65: 229.

Harris, R. S., and Jones, D. M. (1956): The arterial supply to the adult cervical vertebral bodies. *J. Bone Joint Surg.* 38-B: 922.

Hassler, O. (1966): Blood supply to human spinal cord: A microangiographic study. *Arch. Neurol. (Chicago)* 15: 302.

Hassler, O. (1970): The human intervertebral disc: A microangiographical study on its vascular supply at various ages. *Acta Orthop. Scand.* 40: 765.

Henriques, C. Q., and Chir, M. (1962): The veins of the vertebral column and the role in the spread of cancer. *Ann. Royal College of Surgeons of England* 31: 1–16.

Isherwood, I. (1962): Spinal intraosseous venography. *J. Faculty of Radiologists* 13: 73.

Kadyi, H. (1886): Über die Blutgefässe des menschlichen Rückenmarkes. *Anat. Anz.* 1: 304.

Maciver, D. A., and Letts, R. M. (1968): Intraosseous vertebral venography as a diagnostic aid in intervertebral disc disease. *Can. J. Surg.* 11: 16.

MacNab, I., and Dall, D. (1971): The blood supply of the lumbar spine and its application to the technique of inter-transverse lumbar fusion. *J. Bone Joint Surg.* 53-B: 628.

Mannen, T. (1966): Vascular lesions in the spinal cord in the aged: A clinico-pathological study. *Geriatrics* 21: 151.

Nathan, M. H., and Blum, L. (1960): Evaluation of vertebral venography. *Am. J. Roentgenol.* 83: 1027.

Romanes, G. J. (1965): The arterial blood supply of the human spinal cord. *Paraplegia* 2: 199.

Stilwell, D. L. (1959): The vascular supply of vertebral structures (gross anatomy: rabbit and monkey). *Anat. Rec.* 135: 169.

Suh, T. H., and Alexander, L. (1939): Vascular system of the human spinal cord. *Arch. Neurol. Psychiat.* 41: 659.

Tureen, L. L.: Circulation of the spinal cord and the effect of vascular occlusion. *Res. Publ. Assoc. Res. Nerv. Ment. Dis.* 18: 394.

Vogelsang, H. (1970): *Intraosseous Spinal Venography.* Excerpta Medica, Amsterdam.

Wagoner, G. and Pendergrass, E. P. (1932): Intrinsic circulation of the vertebral body with roentgenologic considerations. *Am. J. Roentgenol.* 27: 818.

Willis, T. A. (1949): Nutrient arteries of the vertebral bodies. *J. Bone Joint Surg.* 31-A: 538.

Woolam, D. H. M., and Millen, J. W. (1955): The arterial supply of the spinal cord and its significance. *J. Neurol. Neurosurg. Psychiat.* 18: 97.

A

Abdominal wall branches, of lumbar arteries, 5

Adamkiewicz' artery, 27, 57

Anterior internal vertebral venous plexus, 45, 51–53, 86, 89

Anterior median longitudinal arterial trunk, 25–27, 29–33, 36, 40, 41, 57, 102–105, 109

Anterior median longitudinal venous trunk, 57, 58

Anterior median spinal veins, 103

Anterior radicular artery, 36, 37

Anterior spinal artery, 3, 24, 25

Anterior spinal canal branches, of lumbar arteries, 5

Anteroposterior venous anastomotic channel, 112, 113

Aorta, 12
    injections through, 119
    intercostal arteries from, 4

Arcade system, 45
    of lumbar artery branches, 5

Arcuate branches, of lumbar artery, 17–19, 66, 67

Arcuate arterial pattern
    of spinal canal, 3, 10, 16, 21, 71
    of venous plexus, 45, 51

Arterial grid, in vertebral body, 66

Arteries
    supplying meninges and spinal cord, 24–31
    supplying vertebral column, 2–21
    within spinal cord, 102–117
    within vertebrae, 66–83

Ascending cervical artery, 24

Ascending lumbar veins, 45

Atlantoaxial joint, 8

Atlantoaxial junction, 7

Atlas
    arterial chains of, 2
    facet-joint arteries of, 81, 82
Azygos veins, 44, 45
    injections through, method, 119
    retrograde filling of, 54

B

Basivertebral veins, 86, 88, 89, 90, 91, 92, 93, 98, 99
Brachiocephalic veins, 44

C

Cancellous bone, 66
Cartilage caps, arteries of, 67, 74, 76
Cauda equina, 39, 58, 61, 62
    arteries in, 109
Central arteries, 32
    of spinal cord, 103, 104, 106, 107, 117
Central veins, 111, 117
Centrodorsolateral venous anastomoses, 103, 112, 114
Centrum, basivertebral system of veins of, 86, 88
Centrum branches, of lumbar arteries, 4
Centrum grid, arteries of, 80, 109
Cerebellar arteries, 25
Cervical arteries, 3, 11, 24
Cervical nerves, blood supply of, 3
Cervical veins, 44
Cervical vertebrae, arteries of, 2-3, 7, 8, 67, 80, 81, 83
C2 cervical vertebra, vertebral arteries of, 2, 3
C6 cervical vertebra, vertebral arteries of, 2
Cervicothoracic junction, 9, 11
Conus medullaris, 39, 40, 41, 58, 61, 62
Costocervical trunk, 3
    arteries derived from, 2, 11

D

Deep cervical artery, 24
Descending aorta, 9
Disc, venous network of, 97
Dommisse's nomenclature, for spinal cord arteries, 25
Dorsal median longitudinal venous trunk, 54
Dura, as possible vein valve, 63
Dural sac, 38
    blood supply of, 24, 28, 35

E

Epoxy resin, vein injection with, 51, 119
External vertebral venous plexus, 44, 45, 47

H

Horizontal subarticular collecting vein system, 86, 94, 96, 98

I

Iliolumbar arteries, 5
Inferior thyroid artery, 2
Inferior vena cava, 45
Intercostal arteries, 2, 3, 4, 9, 11, 24, 35
Intercostal veins, 44
Internal oblique muscle, 5
Internal vertebral venous plexus, 44, 45, 51, 62, 63, 75, 87, 99
  description of, 45
Intervertebral foramina, 3, 5, 15, 16, 17, 44, 45, 51, 78
Intervertebral veins, 53
Intraosseous distribution, of arteries, 73, 78, 79
Intraosseous tributaries, of lumbar arteries, 16
Intraosseous veins, 86

J

Jugular veins, obstruction of, 45

L

Laminae, 15, 19
  blood-vessel network over, 6
Laminal arch, blood supply of, 80
Laminar arteries, 3
Latex rubber, vein injection by, 52, 119
Ligamenta flava, blood-vessel network over, 6
Liquid nitrogen, vertebral column snap-freezing by, 119
Longitudinal ligament, 10
  arteries under, 3
Longus colli muscle
  arterial chains of, 2, 7
  arterial plexus in, 3
Lumbar azygos veins, 45
Lumbar arteries, 2, 4, 7, 12, 14–18, 24, 37, 66, 69
  anterior branches of, 5
  anterior spinal canal branches of, 5
  ascending and descending branches of, 4, 67, 70, 77, 90

centrum branches of, 4
origins of, 13
posterior branches of, 5, 6
spinal canal branches of, 5–6, 71, 77
Lumbar nerve, 17
Lumbar plexus, 5
Lumbar spine, blood supply of, 20
Lumbar veins, 45, 46
Lumbar vertebrae
arteries within, 66–83
blood supply of, 3, 4–6, 14
veins of, 46, 48–50, 91
L2 lumbar vertebra, lumbar artery of, 37
Lumbosacral junction, 14
arteries of, 76

M

Marrow spaces, of vertebral body, venous drainage of, 91
Median anteroposterior venous channel, 114
Median sacral artery, 4–5
Median longitudinal venous trunks, 54, 57–59
Median veins, 103, 115, 116, 117
Medulla oblongata, 25
Medullary feeders, 26
measurements of, 27
Meningeal plexus, 24
Meninges, blood supply of, 24–41
Methods, of vein and artery preparation, 119–120
Micropaque
for vein injection, 53, 56, 57, 59, 89, 92, 93
method, 119, 120
Midlumbar spine, blood supply of, 19

N

Nerve root arteries, 24, 26, 32
Nervous system branches
of lumbar arteries, 5–6
of vertebral arteries, 3
Neural branches, of lumbar artery, 17
Newborn, vertebral-body venous systems of, 87, 88

P

Paraspinal muscles, blood vessels in, 6
Pars interarticularis, blood vessels crossing, 6, 19, 20
Pedicle veins, 53
Pelvic plexus, 45

Peritoneum, blood supply of, 4
Pial plexus, 25, 102, 115
Pial plexus artery, 32, 105
Posterior intercostal arteries, 2, 4, 9
Posterior internal vertebral venous plexus, 45, 53
Posterior longitudinal venous trunk, 61, 62
Posterior median longitudinal venous trunks, 103
Posterior median vein, 111
Posterior spinal arteries, 3, 24
Posterolateral longitudinal arterial trunks, of spinal cord, 25,
    26, 27, 30, 33, 34, 35, 39, 40, 41, 59, 61, 102, 109
Prostatic plexus, 45
Psoas muscle, lumbar arteries in, 4, 5

Q

Quadratus lumborum, 5

R

Radiate arteries, 86
Radiate veins, 103, 115, 117
Radicular arteries, 26–27, 30, 32, 36
Radicular vein, 52, 63
Radiculomedullary arteries, 27, 29, 30
Retrograde filling, of veins, through azygos system, 54
Retroperitoneal tissues, blood supply of, 4

S

Sacral arteries, 13, 14, 24, 26
  origin of, 4–5
Sacral vertebrae, arteries of, 67, 79
Sacrospinalis muscles, blood vessels in, 6
Segmental arteries, 26–27
Segmental veins, 59
Spalteholz clearing method, 19, 21, 26, 30, 31, 35, 38, 40, 56,
    57, 99, 105–107, 109–112, 114, 120
Sphenoidal clivus, 45
Spinal canal arteries, 24, 25
  origin of, 3
Spinal canal branches, of lumbar artery, 5–6, 17
Spinal cord
  arteries and veins within, 102–117
  blood supply of, 3, 24–41
  segmental arteries supplying, 26–27, 36
  surface of, blood supply of, 24–26
  veins of, 56–63
  venous anastomoses of, 114

Spinous process, 15, 46, 78
  vein of, 87
Subchondral postcapillary venous network, of vertebral body,
      87, 95, 96, 97, 98
Superior intercostal arterities, 3, 9, 11
Superior vena cava, 45

T

Thoracic nerve, 3
Thoracic vertebrae
  arteries of, 67, 82
  blood supply of, 3–4, 7
T5 thoracic vertebra, intercostal arteries of, 2
Thyrocervical trunk, 3, 7, 11
  arteries derived from, 2, 26
Thyroid artery, inferior, vertebral branches of, 2
Transversus abdominis, 5

V

Veins
  of spinal cord, 56–63
  within spinal cord, 102–117
  of vertebral body, 86–99
  of vertebral column, 44–53
Vena cavae, 45, 119
Venae comitantes, of vertebral arteries, 44
Venous anastomoses, of spinal cord, 114
Venous pial plexus, 103
Vertebrae, arteries within, 66–83
Vertebral arteries, 3, 7, 8, 10, 26
  origins of, 1, 2
Vertebral bodies
  blood supply of, 2, 3, 7–21
  veins of, 86–99
    schematic drawing, 98
Vertebral column
  arteries of, 1–21
  snap-freezing of, 119
Vertebral end-plates
  arterioles of, 66–67
  cartilage capillary bed of, 87
  venous system of, 86, 87, 91, 92–94, 96, 97
Vertebral veins, 44–53
Vertebral venous plexus, 44

X

X-rays, of vertebral-column specimens, 120